Lake Superior GOLD

An Amateur's Guide to Prospecting in the Lake Superior Region

Jim Dwyer

NORTH STAR PRESS OF ST. CLOUD, INC.

Acknowledgments

I have had great fun writing this book. The research has led me to many people who have a great and intimate knowledge of the geology under their feet. They took time to explain and show what they know to an amateur with an interest. Some of these are professional geologists with companies or governments, some are prospectors who go out with a jeep and equipment. I would especially like to acknowledge the work of Mark O'Brien, geologist, Ontario Ministry of Northern Development and Mining in Thunder Bay, who invited me to the annual geology symposium, read the manuscript draft and sent information. Bob Seasor, Copper Range Copper Company, White Pine, Michigan, gave of his time to talk copper and rocks and to steer me to the A. E. Seaman Museum. Richard Rhuhanen, Information Supervisor with the Minnesota Department of Natural Resources, Minerals Division, let me look at maps, core samples and reports. And also, thank you to the many geologists and prospectors who answered questions and told stories. I also wish to thank Susan Gaudino, Resource Geologist, and John Mason, Resident Geologist of the Ontario Ministry of Northern Development and Mines, for information and assistance. I wish to thank, too, John Heine, Assistant Scientist at the Center for Applied Research and Technology Development at the Natural Resources and Research Institute, for his assistance with map-making.

Library of Congress Cataloging-in-Publication Data

Dwyer, Jim, 1953-
    Lake Superior Gold : an amateur's guide to prospecting in the Lake Superior region / Jim Dwyer.
    96 p. 23 cm.
        Includes bibliographical references.
        ISBN 0-87839-067-7 : $9.95
        1. Gold mines and mining—Superior Lake Region. 2. Prospecting—Superior Lake Region. I. Title.
TN423.A5D89 1992   92-14659
622'.1841'097749—dc20   CIP

The photos on pages 3, 23, 29, 33, 35, 37, 38, 39, 42, 43, 52, 54, and 79 are courtesy of the Ontario Ministry of Northern Development and Mines.

Copyright © 1992 Jim Dwyer

All rights reserved. No part of this book may be reproduced in any form without prior written permission from the publisher.

Printed in the United States of America by Versa Press, Inc., Peoria, Illinois.

Published by North Star Press of St. Cloud, Inc., P.O. Box 451, St. Cloud, Minnesota 56302.

ISBN: 0-87839-067-7

Foreword

This book is the result of years of an amateur's interest in geology, prospecting, and human history. As a youth, I learned much while poking around Lake Superior with my father. We would explore around the Iron Range of northern Minnesota, the Keweenaw Peninsula in Upper Michigan or the Sibley Peninsula near Thunder Bay. Now my children join me looking for ore samples, old buildings and evidence of people who have passed by. We have found gold and silver ores, minor gems, and artifacts. The outdoors, fresh air, museums, and pleasant company are all part of our mineral and history hunts.

The Lake Superior region is beautiful with hundreds of beaches and cliffs, campgrounds and parks. There are cities and resorts, open country, birds and bears to be enjoyed. Within two hundred miles north, west or south of Lake Superior, eager prospectors may find gold, silver, gems, ores, and historical relics. Take the time to look. The fun and pleasure of standing in a river panning gold or walking out in an open gravel pit picking rocks adds excitement to a camping trip. I strongly recommend that readers of this book try to interest their families and friends in joining them for a small search. Who knows what you might find in the great Northwoods! I do know that prospecting can add much pleasure to your vacation.

Contents

Introduction v

Lake Superior Treasures 1

Gold and the Geology of the Lake Superior Region 21

Ontario Gold Mining and Prospecting 31

Gold Prospecting in Minnesota 45

Prospecting in Michigan and Wisconsin 57

Silver, Diamonds, and Semi-precious Gems, Copper,
    Ghost Towns, Petrified Wood, Fossils, Concretions,
    and Odd Rocks 65

Sources of Information 81

Glossary 87

*Grandsons of the mining men, you can see it in your dreams.
Beneath your father's bones still lie the undiscovered seams
of quartzite in a serpentine vein that marks the greatest yield
and along the Midland Railway it's still told how the Rawdon
Hills once were touched by Gold.*

"The Rawdon Hills," by Stan Rogers © 1977
Used with permission of Fogarty's Cove Music

Introduction

There are gold, silver, diamonds and gems waiting to be found just a few hours drive from Chicago, Madison, Minneapolis or Toronto. Gold has been found in Minnesota, Wisconsin, Michigan and especially Ontario over the past hundred years and is being mined right now in Ontario. A new mine is opening in Wisconsin. Copper and silver are being mined in Michigan. Major mining companies are searching throughout northeastern Minnesota for deposits significant enough to begin operations.

Amateur prospecting is being encouraged in Ontario. Minnesota has recently recatalogued all its mineral information, making it more accessible to prospectors; testing, as well as prospecting, is going on in Michigan's Upper Peninsula (the U.P. in local parlance). Valuable minerals and gems lie loose in gravel pits or imbedded in ancient rock waiting to be found. Recent predictions indicate that new gold mining operations will be started soon, and after looking at the latest research in the region, I don't think that prospect is too far-fetched.

You may want to spend your vacation exploring the Lake Superior region and prospecting for gold. This book will tell you how to look for gold, silver and gems; where to find them today; and where they have been found in the past. Written for amateur rockhounds, vacationers, or anyone interested in trying a hand at finding something special, the book can't promise that an amateur prospector on a two-week summer vacation will make enough to buy everybody Christmas presents, but the memories and pleasure of searching will provide as much warmth as a Yule fire.

Make no mistake about it, the possibility of finding gold or silver is real. A couple of prospectors from Ontario told me their eight-year-old grandson found gold on one of their claims near Beardmore, Ontario. A part-time prospector found gold-bearing rock in the roots of a blown-down tree when he was logging northwest of Thunder Bay. A forester told me he kicked up a big piece of silver while walking in woods near Ironwood, Michigan. So you or members of your party may also find gold flakes, silver ore, pure copper or a rough gem.

The search for hidden treasures has always been one of the great fascinations for human beings. I feel it, and I know I am not alone in that feeling. To spend some of your time looking for gold or silver will give you a better understanding of the beautiful Lake Superior region, a special focus to your travels, and, quite possibly, a reward you can hold in your hand. It may not be gold, but other treasures exist if you know where to look, how to look, and for what to look.

Major corporations currently hold exploration permits to look for gold, uranium, diamonds, copper, zinc, molybdenum, beryllium and cobalt. A new gold deposit has been announced near Ladysmith, Wisconsin. Test drilling and prospecting is going on near Beardmore/Geraldton, Ontario; near Marquette, Michigan; near Ladysmith, Wisconsin; even at the edge of the Boundary Waters/Quetico wilderness areas on the Minnesota/Ontario border.

You can be a part of this search, looking for gold and silver to sell or agates and gems to polish for your own enjoyment.

This book is intended to give you a general overview of the prospecting opportunities near Lake Superior. There is a chapter (with map marked with known deposit locations) for each state touching Lake Superior and for the Province of Ontario. Additional reference books are mentioned in chapters as well as listed in the bibliography. Ontario and Minnesota have excellent maps of specific sites and mines for sale. So use those references to find your place to prospect.

Some of the places are now parks, some are working mines, some are on private property and some are impossible to locate exactly any more. Remember that, like other gems and minerals, "gold is found where it's found." Do not just look where others have found something before. Do a little looking into the geology of where you wish to search. Ask local people a few questions. Go to the county museums to see what someone's grandfather found. Your vacation area could hold many sources of information.

Sites in the national forests, in state, provincial and local parks and in private campgrounds make nice base camps. Write or call ahead for advance information about camping, wilderness supplies or needed outdoors equipment. Numerous resorts and private areas even can provide a comfortable enjoyment of the wilderness. At the back of this book I have included addresses for the local Chambers of Commerce. You may wish to write to these folks for specific information.

For each chapter I have recorded some of the rumors that I have heard, included because, in the process of researching this book and exploring the Lake Superior region, many people told me of places they heard that "have just the sort of . . . ." Nothing noteworthy has been found there yet. Or, at least, nothing in the record book. Much of what is known is not written for everyone to see. Every major, minor and fool's-gold rush has been sparked and enhanced by rumors. So I am passing on what I heard.

Cautionary Notes:

1. Most land is owned and cared for by someone or some entity these days. When you want to walk some field, *ask the owner*. If you come across an area staked as a claim, you are not allowed to enter or take samples from that claim without permission of the claim owners. Claim-jumping was not taken lightly by the California 49ers, nor is it today. Most people you meet will be happy to assist you in finding places to prospect or areas to search.

2. When you are in the Voyageurs National Park, remember that all the sites, mines, rocks and artifacts are protected. The 1979 United States Archeological Resources Protection Act provides for fines or imprisonment for people who pick up samples or artifacts. Some parks have brochures describing permitted activities in great detail. I recommend you pick up rule books and look them over carefully. Other parks have specific rules covering digging and prospecting. Be sure to ask about and follow all park rules. A Wisconsin geologist told me that all the members of a rock club were ticketed and fined when

they went without checking rules on a rock-picking field trip to a Wild and Scenic river.

3. When you go to pan for gold in a stream, pan gravel. Do not pan dirt. Dirt will not have any gold in it, and you will only muddy the water. Do not leave holes in the banks of the rivers. Clean up after yourself. I hate coming across garbage left by previous visitors. I know landowners hate it, too. And they have a long memory. So, for your own sake and for the rest of us who may follow your trail, think first, then leave the land as you found it.

4. And finally, a word of caution about the lake country weather. Often, the "lake effect," which is the weather caused by Lake Superior's size and temperature, changes the weather near the lake. If you are near the lake prospecting, and it is cold, foggy and uncomfortable, call inland to a park ranger or listen to the local radio forecasts. Often it can be thirty degrees warmer only five to ten miles away from the Lake. Instead of sitting in cold, damp fog, you and your party could be basking in sunshine only ten minutes away. Be ready and willing to move.

Start dreaming of where you might find "color," the prospector's word for a trace of gold in the gold pan. It's out there. Someone will find it soon. You might be the one to "strike it rich." You can hardly lose by trying.

This map shows the area covered by this book.

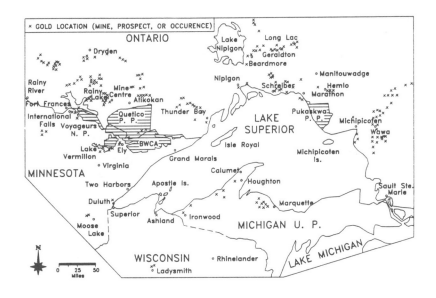

Chapter 1

Lake Superior Treasures

"No gold nuggets? I've found gold nuggets and more in northeastern Minnesota. Not the road work that uncovered Hemlo, but gold," said the lawyer, leaning against the display. "There are other things up here besides gold and silver." "PGE, Platinum Group Elements?" I suggested. He looked at me for a second. "Other things besides gold," he repeated.

Gold, Silver, Gems, Ores, Artifacts and Interesting Rocks.

I think that anyone who goes searching with a serious intent for gold or silver or semi-precious and precious gems in the Lake Superior region will go home rewarded, bringing home rocks bearing gold or silver or a chunk of pure copper. Certainly some beautiful Lake Superior agates with the famous jointed bandings can be found.

Prospecting is encouraged in the Province of Ontario. Vast tracts of public land, "Crown Land," may be explored. Almost everyone in Ontario, it seems, has some stake in a gold claim or mine or prospect. Minnesota wants to hear more about the minerals under the land. Its Department of Natural Resources has been holding mineral lease sales for the past few years, and many companies have bid on the right to prospect in northeastern Minnesota. Wisconsin and Michigan are also willing to help responsible prospectors look for valuable resources under the ground.

Gold is the word everyone wants to hear. Old mines and new locations are being tested in the hope of finding enough metal to warrant the expense of opening operations. Companies are even willing to pay for good information about the location of gold samples.

Gold, a wonderful metal beautiful to look at and wear, is used for money, for protective spaceship coating and for fine electronic parts. Stamped on the spines of books and used in medicines, gold is also an amazing paradox. Unlike iron or copper, gold has no great value to manufacturing. Shiny but too soft for use in machinery, gold's uselessness, combined with its indestructibility and rarity, has made it valuable.

Most of the gold being mined today is not used for jewelry. The majority is held by investors who leave it in bank vaults or government storage. Owning gold shows the world that you have money to salt away. Gold's value on the market reflects the world-wide demand for the metal. Woven into that demand is the history of gold as money, worries about the future of countries, the convertibility of all currencies into gold and an inherent desire for owning gold.

Until recently, the price of gold was set by the United States at twenty dollars an ounce. In Canada, the price of gold varied with market forces but was usually below twenty dollars per ounce. When the United States government allowed the price of gold to be decided by the market place, the price moved immediately to over two hundred dollars an ounce and currently is valued around three hundred fifty dollars (US) per ounce.

Gold is called the ultimate storehouse of value. The reasoning behind this axiom can be simply put this way: in 1892 an ounce of gold could buy your great-grandparent a fine suit of clothes; in 1992 that same ounce will still buy a fine suit of clothes.

Gold, like jewels, has been seen as an investment by some people. This causes the price of gold to fluctuate based on investors' worries about the stock market or the political unrest in the Middle East instead of supply and demand. Gold itself produces no interest or dividends. Money is made by buying the gold cheap and selling when the price goes up.

These same market forces decide whether a mine will open or close. If gold can be produced at, for example, the Hemlo deposits near Marathon, Ontario, at a cost of fifty dollars per quarter ounce then there is a profit to be made as long as the price of gold stays above two hundred dollars an ounce. With the costs of opening a mine increasingly prohibitive, only those deposits which have substantial potential will be developed.

Silver has been mined in much of the region since the mid-1800s. Silver ore is still brought to the surface at some of the copper mines. Closed silver mines are located in downtown Duluth, near Thunder Bay and in the Upper Peninsula of Michigan. Silver is part of many location names in the region, names like Silver Islet, Ontario, and Silver City, Michigan.

Silver Islet, 1921.

Diamonds, emeralds, and garnets have also been found. Lake Superior agates are the best known semi-precious stones found. Other interesting stones include Thomsonite, tiger's eye, amethyst, and other semi-precious stones that will take a polish and make a beautiful ring or necklace.

Copper can be found in many parts of Michigan's Upper Peninsula and in Isle Royale National Park. Float copper is easy to spot by amateurs, beachcombers, even children. Pieces of float copper shine when the tarnish is removed. By adding hoops or chains, beautiful earrings, necklaces or tietacks can be yours.

Many natural stones and minerals are just plain fun to find. Concretions shaped like the Buddha are found near Ironwood, Michigan. Fossils of sharks' teeth, snails, and oysters can be found in the overburden of some iron mines on Minnesota's Iron Range. Petrified wood rests in some of the gravel pits and stone exposures of road work. Even ancient lava can be found. (Information on these items is in Chapter Six).

Fool's Gold and Fraud

A short section must be devoted to iron pyrites or "fool's gold." Iron pyrites form a variation of iron with the color of gold but considerably less density. It can be found in veins, in crystal shapes and in loose gravel. I have found fool's gold in the Busheyhead Island mine drift and on the Iron Range and in most parts of the Lake Superior region.

Fool's gold has been given credit for opening up Minnesota's Iron Ranges. Many claim that fool's gold sparked the gold rush to Lake Vermilion. It gave the miners something to look at until they realized they were standing on massive iron deposits worth the equivalent of many gold mines.

Iron pyrites can be found almost anywhere throughout the Lake country. It frequently makes its appearance embedded as tiny gold flecks in rock. You can see it at times under water or in samples broken open in mining pits. Iron pyrites can usually be determined quickly by the use of a magnet, the scratch test or wetting it and leaving it out overnight to see if it rusts.

Fool's gold has no commercial value as yet. Some researchers are trying to use it as a replacement for real gold in

Fool's gold.

some applications. Regardless, it is fun to find and can be a drawing card to interest children in your prospecting. It's great for Show and Tell at school.

How to Start

If you are already on vacation, use this book to guide you to treasure hunting sites nearby. If you have the time for long-range planning, I recommend some reading tips in the next few pages. Also included at the back of this book is a list of other useful and interesting books and sources of information. Some are special interest books about a particular area and can be bought only near the locations; others must be specially ordered from the publisher.

Reading and planning can add greatly to your chances of success in finding treasure. Just jumping out of the car to pan gold at any stream will probably lead to frustration. Knowing what to look for in a stream in order to find gold will increase your chances of success. This book will describe some of the features of streams and rock formations that prospectors look to see. Other books listed in the bibliography will provide greater details about specific questions.

Some of this research can be done at home during the winter before you begin your vacation. Some will have to wait until you are driving around your vacation area after settling in your base camp or resort. Some research is best left to those inevitable rainy days that go with every vacation.

In order to begin preparations for your trip, consult local

libraries for state or provincial geology texts and maps. At the back of this book are some addresses to write to in order to buy maps or books. Some very good maps and books have very modest prices. These are the same texts and maps used by the mining companies and professional prospectors for their work. You may want to ask companies or agencies for their catalogs of maps and books. Then you can choose those that best suit your needs or locations.

Maps are very useful in orienting you to good prospects. Sources of maps are listed at the back of the book. Some of these maps you can write for in advance. Some you can only see in person. So, if you're reading this while sitting in your hammock near Pukaskwa National Park, you still have time to look at maps.

Copies of some old and new maps also can be found in libraries, tourist centers, historical societies, even in promotions for local events. Most people only give these resources a passing glance. Since you are looking for treasures they may provide significant clues.

Once you have narrowed down where you're going to spend your vacation, call or write to the Natural Resources Department, geological society, or other appropriate ministry or agency. Names and addresses are listed in this book's appendix. Ask for suggestions for specific books, maps and museums or mineral displays within your vacation area.

Historical evidence of prospecting and mining gives another whole list of clues to professional and amateur prospectors. An area where gold was found in the past is often an excellent place to look today. There are many abandoned mines, test pits, and prospects in the region where people have looked for gold, silver or other minerals. With some research or time spent talking to local people, you may come up with a new twist to an old mining site. Remember, many Chinese bosses became very rich in the Old West by using coolie laborers with gold pans to glean the slag piles of played-out gold mines. They found gold by looking at the problem from a different angle.

State and provincial historical societies have information about early settling of the region you will be exploring. Ask what ores and minerals attracted early settlers and miners. Look to see where old towns, settlements, railroad crossings and camps were located. Compare old railroad maps with modern road maps; you will be surprised by how many "towns" no longer exist. Many were boom towns or shanty towns that fell apart when the miners, railroad men, or lumbermen moved on. In some cases settlements were burned over and never rebuilt after the terrible forest fires at the turn of the century swept through. Other towns, camps, and whistle stops just sank back into the earth and forest. These are fun to poke through with a metal detector.

County museums, local historical societies and libraries may have information of interest to you. Information about these is available from towns' Chambers of Commerce. A listing of the Chambers is in the appendix. You will want to ask for information about mining activity, old town sites, and anything else local people might consider important or quirky.

I have found the people who live in the mineral areas very helpful in locating places to prospect, but remember that some areas are private property where prospectors are not wanted, and some sites are already staked and claimed. The No Trespassing signs and the claim stakes will be fairly obvious. Asking permission to prospect or pan is always a good idea when going on private land, but you will cross much public land and many rivers while driving. Use your knowledge to give the best of them a try.

Check old resource files at libraries and newspapers. Periodically, newspapers will run articles or headlines about gold being discovered in some area. It usually pans out to be nothing of great value, but someone at sometime found enough gold or silver or diamonds to make an assayer agree that the rock they held had value. An example I recall is an article from the late sixties about finding gold on the Kettle River in northern Minnesota. Someone brought in samples found in the river that showed gold. Nothing more has been done that I am aware of in that area. Was it just a fluke find in the glacial till? I don't know, but this might be an area worth a look.

Many towns have information booths or interpretive centers. Ask the people working in these establishments. Many volunteers have a strong interest in their local history. They may know local people who have found this or that or whose father or grandparent found something way back when. They may also know who to contact to get permission to look on private land.

Road construction in Duluth exposed new faces of rock for exploration.

Another source of information is the local highway department of the county, state or province. Highway departments usually have outposts sprinkled in the small towns throughout their territory. These outposts often are made up of two to four workers who form that area's patrol, grading roads and doing other upkeep. Often they are at the outpost only early in the morning or late in the afternoon to pick up equipment. The rest of the day they are on the road working. Ask them where area gravel pits and loose rock washouts may be found. Inquire about road construction that might expose new faces in rock walls or gravel-lined hillsides. Ask about rock outcroppings or cliff faces that give road workers trouble. These all may be great sources for rock specimens. One gold prospecting company told me it sends some geologists along all the new road work in northeastern Minnesota to check outcroppings.

Artists often make use of local materials to create artifacts, craft work and trinkets. Asking about an interesting stone or ore sample in an art shop might lead you to where the artist found the raw material. By using your detecting skills, the looking for where to search becomes a real treasure hunt.

How to Prospect

Prospecting is a catch-all term that connotes looking for a mineral deposit. You can prospect with a hammer or pick, with a gold pan, with a metal detector or even with a tree pruner. When prospecting, you are looking for samples of minerals or of rock that bears the minerals desired. In this book, the prospector's goal is to find proof that gold is in the ground underfoot.

Much of the Lake Superior region has bedrock that is exposed at or near the surface. To prospect rock outcroppings or faces, walk over the rock looking for quartz or serpentine veins. When these veins are found, look them over closely for evidence of folding or twisting. Most of the regions' gold has come to the surface along these twisted veins. If there is something that looks interesting, try breaking a three-to-five pound sample loose. Look it over closely under a magnifying glass. If it appears gold may be in the piece, label the rock with the location found and store it away for assaying. If a site looks particularly good, try to get three or more representative samples. Remember, if you are twenty feet away from a different rock type, you might just as well be on another planet.

Prospecting is also done in other ways. Panning for gold is one, using metal detectors and other high-tech tools are others. These are discussed in the next section. Prospecting using the tree-pruner is very real; it is a technique called biogeochemical prospecting. Biogeochemical prospecting uses organic matter like leaves or bark to determine if there are metals in the soil. This works because gold and other minerals are drawn up into living plants along with the nutrients from the soil. If prospectors find gold in plants, it had to come from the ground below.

This form of prospecting involves collecting soil or vegetation samples for assaying. Anyone can use this method. It does require care in collecting and handling specimens and requires advanced assaying techniques. The assaying is not much more expensive than basic fire assaying. Simple tools, good record keeping and persistence may yield useful information. If you have property that does not have outcrops or bedrock, this method may disclose the existence of gold in the glacial till or the bedrock.

To start biogeochemical prospecting, look over the land to

be prospected. Specifically, find plants that are on most parts of the land. Three species that have shown good results for researchers are balsam fir, black spruce, and Labrador tea. Mark locations across the property where you take samples. On a forty-acre parcel, six samples would be the minimum. Researchers may take samples every one hundred feet in any direction.

Remove any gold rings or other jewelry. Minute scrapings from these may severely skew the results of assaying. Then collect samples from the trees or plants, looking for seven years or more of growth if possible. On balsams, count back seven growth buds on branches and clip about three pounds per location. On black spruce, cut off the scaly dead bark. With Labrador tea, cut the above-ground part of the plant. In all cases, be consistent. Take these to a work area and carefully clean branches of all leaves or needles. In general, branch material is a more representative sampling medium.

Call an assaying firm and ask what kind of sample preparation you need to do at this point. Most will have you send the samples. They will ash the samples and scan the ash, analyzing for minerals. A reading of two parts-per-billion (PPB) gold is considered worth following up. With this small reading, it is easy to see why a gold wedding band could skew the data.

If good news is received from the assaying company, pin point the area where the sample occurred and repeat the process with sample sites much closer together. Keep in mind that the level of metals within living plant material can vary considerably over the course of a year. Resampling should be done as quickly as possible to minimize this seasonal variation. Detailed information about this process may be obtained from the Center for Applied Research and Technology Development. The address is listed in the appendix.

What to Take

After deciding where to hunt for gold, silver or various other metals, mined ores and gems, the tools you bring along for prospecting will depend on how serious you are in your search. Following are three lists of tools and items that you might consider. If you are only going to check a few streams and beaches near where you are staying, look in the first section. The more you are going to do, the farther down the list of tools you should go. All of these tools are available at major hardware stores in the region. Some you may already have in your garage.

For a few hours of looking, take:

a day pack
water and lunch
a magnet
a compass
area maps
bug repellant in season
a geologist's hammer (or any hammer)
safety glasses

If you are going to be out for a longer search, take all of the above and start adding the following:

a magnifying glass
a gold pan
a few bottles to hold samples
a waterproof pen for labeling sample bottles
bags to hold the rock samples
a tweezer

For one- or two-week trips take all of the above plus your camping gear and add the following:

chisels
a pry bar
hiking boots
waterproof pouches for maps and matches
a metal detector (may be rented locally)
a mule (optional, but it makes you look authentic)

There is no general agreement about the preference for using a gold pan or a metal detector for prospecting in the Lake Superior region. I feel both are of real value in prospecting certain areas. It is up to you to decide how much money you want to spend on equipment and how much energy to spend on hauling gear. After reading the discussion on the use of the gold pan and metal detector, you can decide if you want to take them out with you.

Sample set of prospecting tools: pack, metal detector, gold pan, safety hat, pick, compass, chisels, hammer, magnifying glass, horseshoe magnet, shovel, pry bars. All these tools are easily packed and carried by one person.

Gold Pans

Panning for gold is a standard method for testing whether there are sufficient gold deposits to warrant further prospecting in the area. After panning for and finding gold you work upstream looking for larger deposits. You do this until suddenly the gold no longer shows up in your pan. This would indicate that you have passed by the spot (called "the mother lode") where the gold washed out of the vein in the rock. So you go back downstream and look for a likely digging spot for the mine. Sounds simple, but first there has to be gold. And while panning is useful in streams, it has limitations when used on the lakeshore.

The gold fields around Lake Superior are geologically much older and much more weathered than California or Alaska where the gold pan was so common. The gold and other metals which were once intruded into or precipitated into the ancient mountains around Superior have long since been weathered out. Gold veins and concentrations are now often covered by glacial till and topsoil. You need to look in gravel pits or gravel river banks where gold may be mixed with glacial debris. Some gravel prospecting is being done in Minnesota. I discuss this in the chapter on Minnesota gold prospecting operations.

The gold pan is bulky but very necessary if you are going to be where there is gravel to prospect. The pan helps you wash

Lake Superior Treasures / 13

Gold pan and rock hammer.

away everything but the heaviest metals that are in the gravel. One of the heaviest metals is gold. You could accomplish the same results by using a tweezer to examine every particle on your shovel full of dirt. That just is not very much fun, and there's no romance in saying, "I spent two weeks tweezing for gold."

The gold pan is a special tool that you might want to buy. Pans can be bought at many of the "real" hardware stores in the region and come in different sizes with prices between seven and twenty dollars. The larger sizes, fourteen- and sixteen-inch pans, are more practical simply for the additional volume of gravel they wash. You may want a smaller pan if your arms do not have a lot of strength or if you will be backpacking into an area you are not sure even has gravel. Smaller pans are also easier for children to use.

The pan should be used only for gold prospecting. Do not use it for cooking soup or holding nuts and bolts all winter. This is not because you worship the pan. Fat and food particles or the dents from nuts and bolts ruin the pan's washing effective-

Panning for gold in a stream.

ness. The gold in the pan usually will be visible only with a magnifying lens. So do not do things to interfere with the pan's natural ability to hold onto tiny grains.

Gold panning works on the simple premise that heavy minerals and metals sink faster in water than lighter material and dirt. Gold is heavier than most everything you might pick out of the ground. So it sinks along with other heavy substances like iron magnetite, pyrite or ilmenite. Together, these are usually referred to as "black sand," and that is just what they look like in the bottom of your washed pan of gravel. When gold is found in a pan, it is almost always found mixed with this black sand. But just because you have black sand does not mean you have gold. It does mean you have done a good job of washing the gravel down.

Most gold panning is done by swirling water over the gravel.

You want the water to help you separate the gold from all of the other material. By swirling the water gently around the pan you gradually wash off everything but the heaviest materials. If you swirl too fast, the gold or silver washes out of the pan. If you swirl too slow, you do not separate the gold from the gravel. When your heaping pan of gravel and dirt is washed down to black sand, you have done one pan full of prospecting. There is a method of panning without water, but I have found it difficult and not very productive.

After you have bought your gold pan, practice with it at any nearby stream or beach. Fill up the pan with sand, gravel and stones. Start by picking out the stones, giving each a quick glance for value as you toss it. It would be a shame to toss out a nice agate or a diamond just because you are looking for gold.

Then hold the pan on opposite sides and put the pan in the water. Let the water run over the dirt. Then slowly start to rock the pan so the water swirls around the pan and splashes lightly over the lip of the pan. Keep the pan lip just above the surface of the water so that you can keep adding water as washing water swirls off. You also save a sore back by not lifting a whole pan of watery gravel. You will do this until the water has washed away everything but the small sand at the bottom of the pan. This whole process should take you about five minutes or less as you become more adept with the pan.

Look at the remaining sand with your magnifying glass. If there are some flecks that look like gold, put a little water in the pan and pour all the black sand and gold into a collecting vial. Silver will look like the lead of a bullet, so save that, too. As you work down the pan load, check the small gravel pieces for bits of agate, garnet or other gems that may be worth keeping.

Metal Detectors

The value of metal detectors for amateur prospecting is still open to question. On the plus side, they can quickly detect a chunk of metal within a few inches of the surface or a heavy concentration of ore. They can also find copper pieces among the stones of a beach. On the minus side, most gold and silver veins are so small that a detector will not give a reading, and

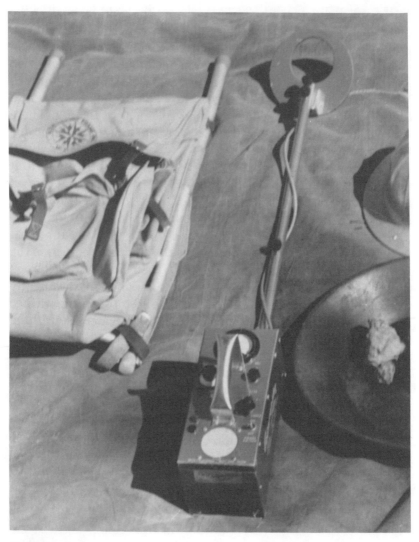

Metal detector.

it will pick up readings from other less-valuable minerals like hematite. It also weighs more to carry. Most prospectors to whom I have talked do not use detectors, but most of the bigger prospecting firms use similar high-tech equipment to aid their prospecting. If you have one or can rent one for a day or two, it may be worth having along.

Gold Field and Rock Hunting Terms

When talking rocks and minerals, it is good to have a working knowledge of a few of the terms that prospectors use to describe their work. It helps to know, for example, if you are looking at a rock or a mineral. And some definitions will help you understand this book.

Rock: A group of minerals that is large enough to be mapped by surveyors or geologists working on the earth's crust. We are talking about large pieces of stone with this word. A small stone may be a piece of rock.

Mineral: A naturally occurring, homogenous solid with a definite (but generally not fixed) chemical makeup in an ordered atomic arrangement. This means, for example, the atoms of gold are all alike to the average eye.

Stone: A piece of rock or mineral that has broken off the rock. The stone may be round and smooth or sharp with edges.

Hardness: The ability of a mineral to resist scratching or to scratch another mineral. Harder minerals scratch softer minerals. You may want to carry along a few known rock specimens like granite, graywacke and talc. Use them to help identify samples by scratching other rocks you find to do some field identification.

Igneous rock: Those rocks formed (or "ignited") under intense heat. Volcanic rock may have come to the surface of the earth or squeezed into existing rock formations. If it came out under water it may have absorbed some of the minerals of the water.

Metamorphic rock: Rock changed because of pressure or heat. Metamorphic rock is very common in the region, because the age of the rock is so great that volcanic and tectonic activities have squeezed, twisted and stretched what we think of as solid rock.

Sedimentary rock: Rock formed when eroded particles of rock bond together because of pressure or heat. Examples of this are sandstone and petroskyite, the Michigan state stone. Sedimentary rocks are the results of millions of years of weathering and reforming.

Conglomerate: A combination of cobble and gravel with sand compressed into a sedimentary rock. Conglomerate can be recognized by the compacted, weathered stones that make up much of its content.

Trenching: Many professional prospectors trench and sample bedrock to find samples worth assaying. Trenching is just what the term implies. All the dirt above the bedrock is removed, exposing faults, shear zones, dikes and folding patterns. Large trenches are dug by a backhoe or bulldozer, making it an expensive and time-consuming effort beyond the scope of amateur prospecting. I include it here because you may come across some trenching work, and it is interesting.

Mine: A location where gold is found in quantities significant enough to be worked. An example is the Ropes Gold Mine near Marquette, Michigan.

Prospect: A location where gold is known to exist and work is being done to determine if mining is feasible. An example is the Raspberry Prospect near Ely, Minnesota.

Occurrence: A location where gold has been found but not in sufficient quantity to warrant the expense of extensive drilling. An example is the Powell Occurrence just east of Quetico Park, Ontario. The goal of the prospector is to move up the prospecting scale from finding an occurrence to opening a mine.

Cautions

Be sure to leave word with someone as to your location and general plans in the event you do not return by nightfall. While wandering in the woods, on back roads and along the shores of Lake Superior, you can become more engrossed in what you see than in where you are.

While looking for places to prospect, do keep an eye out for the artifacts of days gone by. They appear in some strange places. I found a small oil lamp bottle from a wooden sailing ship while looking for rocks along the breakwater at Marquette, Michigan. About to be smashed on the rocks, I grabbed it just in time. A couple of friends found an old earthenware jug floating in the Cloquet River while they were canoeing. They think it washed loose from an old logging camp along the shores.

And if you are going to just stick your find in the basement until you throw it away, consider bringing it to the local historical museum near where you found it. They may even know who once used what you now hold.

Islands should not be overlooked when prospecting. It may be that the minerals help stiffen the surrounding rock formation, or it may be just pure chance, but a number of mines are located on islands. Something in the geologic structure of the islands in this area perhaps encourages veins of gold and silver to show on them. Gold was found on Busheyhead Island in Rainy Lake. Silver Islet near Thunder Bay had wonderful deposits of silver. On Isle Royale there are remains of ancient pure copper mines. A stop on an island may prove pleasurable and profitable.

Assaying

Assaying is the process of separating minerals from host rock to determine the content. You will probably not be able to do this yourself. Assaying is an inexpensive process that you can have done by simply sending your ore samples through the mail. In the back of the book, I have included a few names of assaying companies that will test your samples for about $20 to $50 per four-pound sample. Call or write them to get their latest prices, services and requirements before you send samples.

The most common method of assaying is fire assaying. To do this, the assayer crushes the rock into a very fine powder, and, simplistically, heats this until any gold, silver, copper or other metal "melts out into a puddle." Fire assaying works because metals melt at different temperatures; gold drips out of a sample at one temperature and silver at another. The metal is then cooled and weighed. The measured results are used to determine if the sample has any dollar value. Samples and assayed metal can be returned to you along with your report.

More exotic methods of assaying are available at higher costs. If you believe your samples warrant the cost, ask for descriptions of special services. One includes using X-rays to look for special halos that would indicate gold. Another uses atomic absorption to determine element content.

If the assay results are very positive, you have to ask yourself, "Who should I tell about what I found?" This is tricky because you will have obligations to the land owner and the government. Other prospectors and companies will want to look at your findings. They may stake claims before you or next to your find. But those questions are beyond the scope of this book. At that point you will want professional advice before proceeding. First, you have to find the gold.

The cash value of your find may not be great. But the pleasure of a warm afternoon poking through a gravel pit or on the shores of sparkling Lake Superior is worth much in its own right. Perhaps you will spend a couple of hours alone or with a friend poking through what was once a thriving lumber camp tucked away in woods accessible only by canoe. Perhaps your family will spend three days playing in the white sands of White Fish Bay while you investigate the mining history to the west. You will bring home wonderful memories and an intimate knowledge of some corner of Lake Superior.

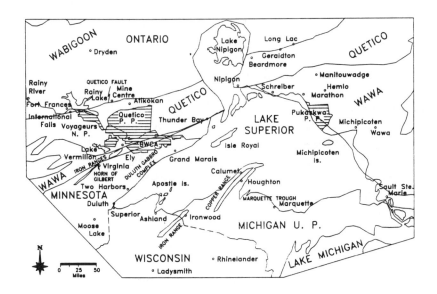

Chapter 2
Gold and the Geology of the Lake Superior Region

"Now, imagine these two rooms are slabs of rock," the company geologist said. "This wall here is between the two rooms. Now the formation I am describing is like these two rooms with the wall being a dike intrusion. The trouble with the analogy is that one room is forty-three-million years older than the other. Apparently they couldn't get their E.I.S. (Environmental Impact Study) approved either."

While driving the Lake Superior region, you will see many exposed rock faces, old mine sites and gravel streams. The question is how to tell if the rock you see might bear gold or silver or copper. To know where to begin looking, it helps to know why gold and other ores are here. This chapter will be a simple overview of the geological history of the area and what is known about gold deposits that have been found.

This area has had much volcanic activity over the past two plus billion years. With all the geological changes that occurred, geologists are unable to agree on how gold came to be in this region. I am aware of two theories. One is that gold precipitated out of lava-derived solutions when the lava touched seawater. This is a sound theory, assuming that when volcanic activity in the region occurred, it happened under the ocean. The rock that spewed forth separated quickly because of the cooling effect of the ocean. Upon precipitation, the gold may have become enclosed within pyrite grains. Certain iron-bearing rock have atomic structures that seem to attract gold. This whole region has much iron-bearing rock and some of it may have come to the surface with the lava and attracted gold.

The second theory is that hydro-thermal forces, like the steam at Old Faithful Geyser in Yellowstone National Park, forced gold silicates through the rock. These hydro-thermal forces held gold in suspension. When the steam passed through the rock, the gold may have been trapped by tiny cracks called micro-fissures—in the rock. Only special forms of the iron molecule become associated with gold because, near many of the prospects, much iron-bearing rock does not have gold.

The portions of Michigan, Minnesota, Ontario, and Wisconsin that this book covers have a similar geologic history. They have the same Precambrian rock (two to three billion years old) that hosts the ore for the gold, silver and uranium mines of the Canadian Shield. Every part of the region was subject to similar variations that occurred under the earth's surface. Molten lava, tectonic shifts and massive earthquakes have all been part of this history. Geological events include the glaciers that so recently shaped the landscape. Even humans, in their very short time here in the region have dug huge pits, blasted and exposed new rock faces and piled growing hills of tailings.

Geologists do agree that the Lake Superior region was the scene of extensive volcanic and plate tectonic activity long before the dinosaurs, let alone motor homes. Rock twisted and stretched, was flooded and eroded, sank into the earth and came back to the surface again, perhaps many times. For example, a stone may have started as lava, was sheared by plate tectonics, twisted into metamorphic rock, eroded into beach sand, reformed into sandstone, was intruded by quartz, sheared again, scraped by a glacier and then dumped in a heap by a copper mine conveyor belt.

The volcanic activity of the region was significant both above and below ground. Volcanoes poured layers upon other layers of lava on the region. Intrusions of granite and extrusions of basalt raised the ground level. In some cases lava flows reached a thickness greater than two miles. The weight of this warm lava caused sinking of the land in some places, while other land rose. The up and down motion twisted some rock structures into rope shapes you can still see today.

All this volcanic activity took place over millions of years. It is not unusual to find two different kinds of rock side-by-side that are fifty million years apart in age. That is why gold may be found in one type of rock while ten feet away there is no trace of gold.

Plate tectonics is the moving about of large pieces of Earth. These pieces, or plates are huge, some the size of half the United States. Plates crash into or scrape along each other very slowly but with tremendous force. The force causes one plate to be forced downward into the ground and the other upward. This

"The Pillows," an interesting geological feature at Poplar Point, Lake Nipigon, formed by lava flows.

is the source of the pressure forcing sediment to bond into rock. The deeper a plate of rock is pushed, the more intense the heat applied to it by the Earth. Such great forces of heat and pressure can stretch, twist and fold rock until its original rock material cannot be determined.

Michigan's Keweenaw Peninsula was an area of extensive plate tectonic movements two billion years ago. Two major tectonic plates slowly crushed into each other. This movement, similar to the faults of California today, pushed surface rock down to a depth of almost four miles. Keweenawan lava extrusions of basaltic material resulted in many of the interesting minerals and rock specimens found along the North Shore. It is in this basalt that quartzite and agates can be found. If you inspect rock outcroppings, you may easily see visible veins of quartzite and similar intrusions.

As plates slide against each other, weak areas—called faults—are created. These fault lines often serve as conduits for lava, steam and minerals to squeeze up to the surface. Along the fault lines of the greenstone belts in the Lake Superior region is where most of the gold activity occurs. The greenstone belts are described later in the chapter.

Oceans may have covered the whole region as some of these basalt intrusions occurred. Most of the bedrock may have been formed under water, much like the Hawaiian Islands are forming today. Some theorists say that the minerals of the area may have come from the seawater. Later, oceans deposited their life-forms remains on the ocean floor and then receded. That is why you may find marine fossils on St. Joseph Island, Ontario, and fossilized sharks' teeth in northern Minnesota and petrified coral colonies in Petoskey, Michigan.

The glaciers changed the topography of the region with their repeated grinding and crushing. Glaciers cracked rock, plucked up the pieces and slowly moved them down the continent. Glacial activity also transported pieces bearing gold, silver, uranium and whatever else was in the way, carrying them where the ice went. When the thaw came, the melting ice deposited the rocks and debris where the warmth found them.

This glacial debris is known as glacial till, defining a confused jumble of dirt, sand, and stones that may have traveled a few feet or many miles. It can be a few feet deep or hundreds of feet thick, depending on the melting patterns of the glacier that dropped it.

Sometimes the melting process sorted and stratified the sand, gravel and rock. These glacial features are called eskers, kames, deltas or such according to their shape and method of forming. Most of the gravel pits of the region are excavations of glacial tills. Looking there takes advantage of some sorting done for you by the glacier. Of course, it may have been undone by the gravel pit operators.

Searching through glacial till in a gravel pit can be rewarding and frustrating at the same time. For example, I found pieces of petrified wood in a gravel pit near Duluth. But they were not big enough to identify the species of trees nor indicate where the trees originally grew.

Topsoil—the organic remains of thousands of years of leaves, trees, grasses, and other organic material—covered whatever the glaciers left behind. In this region, topsoil forms at the rate of about one inch per one thousand years. So, ten inches is the general depth of topsoil. In some areas, erosion, fire, or excavation by humans have left thin (or no) topsoil.

Major Geological Formations to Explore

Certain areas of the region are well known because of their overall size or their importance in mineral production. This is not an extensive description and does not list some minor formations.

Greenstone formations

Greenstone is a basalt rock extruded into the area during volcanic activity. Its greenish color is due to the mineral chlorite. Greenstone formations or "belts" have been found and mapped in most of the Lake Superior region. They are called belts because they are usually long and narrow.

When looking at geological maps, the greenstone belts are usually outlined and named. For example, the Ely Greenstone Belt (Minnesota) contains some of the oldest known rock in the world. It has been dated in this region to approximately three billion years of age. The Ely belt extends across part of northern Minnesota into Ontario.

Almost all the major gold activity has occurred in greenstone belts. Gold is usually found where the edge of the greenstone formation meets different rock. Faults at the meeting of the rock seem to serve as conduits for gold to squeeze to the surface. Basement fractures with microscopic cracks may have acted as sieves to trap gold particles. This may be the case with the Hemlo Deposit near Marathon, Ontario.

In the Beardmore-Geraldton, Ontario area, greenstone intrusions have been the source of major gold finds in the past. Greenstone formations are being worked today by many prospectors and mining firms. The towns of Beardmore and Geraldton were created just to service the gold mining activity. After the gold played out in the 1950s, the towns declined. The mines of the area are described in greater detail in the chapter on Ontario mining. Millions of dollars of gold have been taken out of these formations.

"Horns," or areas where greenstone has been twisted and pulled into "S" shapes, seem to be particularly rewarding when looking for gold. These may be where the intense pressures of volcanoes found release from under the massive rock. The best mines in the Beardmore-Geraldton area of Ontario were located in horns. I watched a geologist from a prospecting group in Thunder Bay try to interest a potentional Quebec investor in putting money up for more exploration of another horn near Geraldton. The Virginia Horn (Minnesota) is actively being drilled, even in some people's back yards. So, look to the ends and edges of greenstone belts when searching maps for your prospecting locations.

Gabbro

The Duluth Gabbro Complex is one of the major rock formations underlying northeastern Minnesota. Gabbro is a dark, dense, igneous rock, a host for copper and nickel-bearing minerals. The Duluth Gabbro Complex is old in geological time. It is a major intrusion that probably resulted from tectonic action along the southwest arm of the Superior Rift. The Sawtooth Mountains near Grand Marais, Minnesota, are the result of gabbro shifting due to tectonic forces. Samples of gabbro can be found along outcroppings, in highway cuts and on the beaches of Lake Superior. Old gold and silver mines in this rock are described later.

The first copper mining efforts to pay off were dug in the location of the old pits near Ontanagon and Rockland, Michigan.

Iron Ranges and the Copper Range

The famous iron ranges of the region have provided millions of dollars of ore. In northeastern Minnesota the Cayuna, Mesabi and Vermilion ranges stretch in long narrow bands towards the northeast. Near Hurley, in northern Wisconsin, the Gogebic Range almost parallels the Minnesota ranges. The Gogebic Range reaches into Michigan near the town of Ironwood. Further east in Michigan, from Iron River to Norway, lies the Menomonee Range. The iron in these ranges probably collected as the result of erosion and settled in Precambrian time. In Minnesota the ore lay close to the surface, and most was strip mined. The Gogebic and Menomonee ranges were mined with shafts. The maps show where to look for fossils at mine sites.

The Copper Range runs from the area around Victoria and Rockland, Michigan, up to the very tip of the Keweenaw Peninsula at Copper Harbor. This area was rich with massive native lodes of copper-bearing ores and copper veins. Some of the copper-bearing rock surfaced on Isle Royale and near Mamainse. The great copper deposits were formed largely by the volcanic and tectonic activities of the region. (More on copper in Chapter Six.)

The histories of these ranges are well documented in the

museums and interpretative centers located throughout the region. It would be well worth the time spent to do some background research as well as view some of the ore samples available at these centers. Addresses are listed in the appendix of this book.

The Quetico Fault

The Quetico Fault runs east-west from Rainy Lake to Dorion, Ontario. This is a major dormant fault line with much mining activity along its length. Nickel-copper mining has taken place along it, as well as exploration for platinum group elements, called PGE on the maps. Old gold mines exist in this fault area. A group of silver mines is found near Mine Centre. These are discussed in the chapter on Ontario.

The Dead River Basin

The Dead River Basin is the area of the only working gold mine in northern Michigan. The Basin is northwest from Marquette. Its shape can be mapped from the edge of Lake Superior to below the Huron Mountains to south by Ishpeming. Within this area is the Marquette Trough, the rock worked by the Ropes Gold Mine. In the past, attempts have been made to establish other gold mines.

What to Look for in These Formations

To find gold in rock country, look for outcroppings of bedrock. These may be domes in the woods or new road cuts for railroads and highways. Look for quartz veins that twist and fold. Chip out some samples and crack them open with your hammer. Examine the veins with your magnifying glass for the glint of gold or silver. If a glint catches your eye, label the specimen so you can tell where it was found.

In loose gravel and sand, gold looks just like the gold in rings in jewelry stores. It is easy to see with a magnifying glass. Look among the roots of trees blown down by the wind. One prospector told me that, while hauling wood, he took a sample from a blowdown that showed color.

As you travel through the area, you can see gravel pits along the roadsides, convoluted rock outcroppings, and tree-covered piles of exhausted ore. Look to see patterns in the scratchings (called striae) on outcroppings so that you can determine the glacier's direction of travel. Watch for crystal faces that glint from a particular mineral and indicate weak routes in the bedrock. Note the strata lines in exposed but undisturbed gravel in pits. All these give you clues about where to look. They make the countryside come alive with possibilities.

A twisted quartz vein in rock.

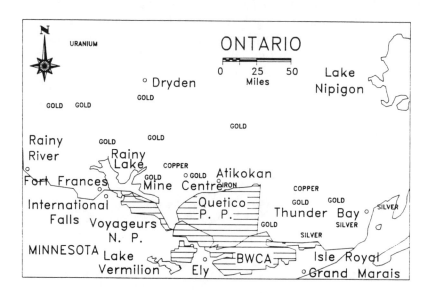

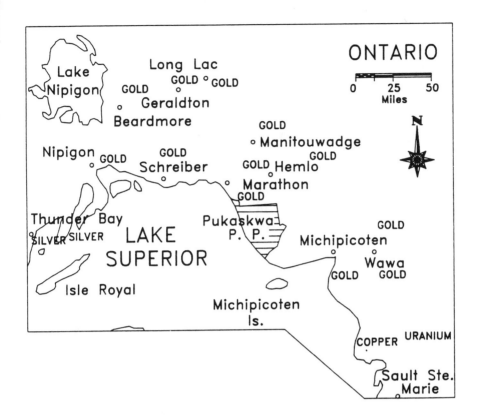

Chapter 3
Ontario Gold Mining and Prospecting

At the reception, the wiry prospector held a bottle of Labatt's Blue beer in nicotine-stained fingers and grinned ruefully at me. "I drove a D-9, cutting trenches at Hemlo. Before, I didn't know what was down under there." He kind of grinned again and walked away, first tossing a glance at the evening's speaker, the man who had found the gold.

Some prospecting and mining now going on in the Lake Superior region occurs on very back stretches of deep woods. Some is going on right in plain view. You may drive right by a drilling rig. You may find it fun to stop and ask a few questions. Drillers may give you evasive answers about what they have found, but I think you will get good information about the equipment or the prospecting.

I have talked with prospectors, company geologists, and government experts about their work and their efforts. I have poured over maps and reports, looking at the fascinating details as detective's clues. And it is detective work. You have clues to sort and sift, hearsay evidence to weigh, historical data to review and your own intuition to follow to find that place that will show color.

Within two hundred miles or so of Lake Superior, a lot of gold activity occurred in the past. Along the Canadian National Railway lines, gold was first found, probably because the road beds opened the area for prospecting. The famous Geraldton Deposits produced thousands of ounces of gold in a forty-year life span. Near Atikokan there are the remains of mines and mining buildings. The Hemlo Deposit mining companies now trade on the New York Stock Exchange. In Minnesota, there was a gold rush. Gold (or the rumor of gold) was one of the reasons the United States bought the Upper Peninsula of Michigan. In Wisconsin, there has been gold associated with the iron ranges.

Ontario encourages responsible prospecting. The Ministry of Northern Development and Mines will provide maps, geological data and prospecting courses for those searching for that big find. They have a Resident Geologist program that puts trained staff people into every mining region to help coordinate information. (I have listed some addresses of these in the appendix.) Ontario even provides grant money that prospectors may use to pay for drilling likely possibilities. Ontario knows that it makes good business sense to have independent, interested prospectors combing the country.

In 1989, over eleven thousand mining claims were on active status in the Thunder Bay area with twenty-nine being advanced exploration efforts. The Ontario Ministry of Northern Development and Mines reported 193 million dollars (Canadian) spent for gold exploration in 1987. Hard-headed business people do not spend money looking for gold if there is no hope for return of their money. They all look for another Hemlo deposit. You might as well look, too.

The Hemlo Deposit

Prospectors have found many gold sites north of Lake Superior including the biggest gold deposit to date, the Hemlo Deposit, near Marathon, Ontario. It is the major gold find of the Lake Superior region both in terms of the size of the mining operation and the amount of gold under the ground. It is estimated that the Hemlo Deposit holds a block of gold-bearing ore worth nearly eight billion dollars (US). This find was not really struck until May of 1981. People had been searching for gold in the area since 1869. However, the crucial hole was not drilled until recently. The concentration of gold in the rock is very small, about one quarter ounce per ton of ore. This means that four tons of ore must be ground up for every ounce of gold extracted. This small amount is feasible to mine with the price of gold currently near four hundred dollars per ounce. If that small amount of gold excites investors, anything you find will certainly be of interest.

Three mines are working the Hemlo Deposit: the Williams, the Golden Giant, and the David Bell. These three mines ac-

The three mines and plants of the Hemlo Gold Mine. In the front is the David Bell Mine; the middle complex is the Golden Giant Mine; and in the distance is the Williams Mine. The TransCanada Highway borders the Hemlo on the left.

counted for one-half of all the gold produced in Ontario in 1989. The Williams Mine, owned by Corona Resources, brought up 495,000 ounces of gold in 1989. The same year the Golden Giant produced 384,000 ounces, and the David Bell 320,000 ounces. That is more gold than most mines produce in their entire life span. The three mines together produced 1.5% of the world's gold output in 1989.

The history of how these three mines came into being and who owns them is long and involved. The story includes gold being found in 1869 by an Indian traveling near Heron Bay. In the 1940s, gold was found in an outcrop in the Hemlo area. A Dr. Jock Williams, amateur geologist, staked some patented claims that later proved very valuable. Surrounding land was staked; drilling and testing were done. In 1947, the test drilling stopped. It would later be discovered that they were just a little above the mother lode.

Most claims were allowed to lapse, except for the Williams claim, but no activity occurred there until test drilling began again in the middle 1970s. Again, the results were disappointing to the companies involved. Finally, some geologists, including a Mr. David Bell, hit upon the mother lode. Suddenly great interest grew in who owned which claims. Litigation went on for years. The Supreme Court of Canada determined who owned this huge fortune. The names of the mines reflect some of this heritage.

Hemlo Deposit is found on a fault line that shows the effects of intense shearing, folding and deformation. The rock on either side of the fault line is about fifty million years different in age. This means that it could be very likely that the gold came up through the weak area of the fault. Or, according to the alternate theory, the rock on the one side was laid down under the water and the gold precipitated out and attached itself to the iron composition. The geologists that I talked with who work on the Hemlo Deposit did not completely agree with either theory. Whichever theory is correct, Hemlo proved a lot of gold and fueled new exploration and prospecting interests.

Wawa, Manitouwadge and Schreiber

The gold fields to the northeast of Lake Superior include the mines near Manitouwadge, Wawa and Schreiber. Near the

Hemlo Deposit is Heron Bay. This area is again going to be the site of test drilling in the 1990s. A new company took over the rights to prospect there, and they are anxious to drill. The Heron Islands also have some drill rigs working. They belong to Teck-Corona Corporation, owners of the Williams and David Bell mines. You may be able to see the equipment from the shoreline.

North of Sault Ste. Marie and east of Lake Superior is the Wawa gold area. Gold occurrences surround Wawa with most occurrences and mines to the northeast. Look for references to the Michipicoten Gold Mine, the Edwards Gold Mine, the Cline Lake Gold mine and others. Iron mines and occurrences are also found near Wawa, as is the Geco copper mine. The Geco also produces zinc, gold, and silver.

The Michipicoten area, near Wawa, has seen much prospecting and mining for gold. One story tells of an Indian couple finding gold in the rock on the shoreline and starting the rush. Early mining did not give good results, however, and the area remained quiet until the 1920s when the technology and milling techniques improved. Between 1930 and 1939 the Grace, Jubilee, Minto and Parkhill mines produced gold and silver in excess of three million dollars.

North and west of the Hemlo Deposit, near the town of Marathon, significant historical and present day activity is in evidence. Near Manitouwadge Lake, four mines brought up

The North Shores Gold Mine near Wawa, Ontario.

gold and silver as by-products to their copper, zinc and lead mining. Some of the old workings and pits can be seen in the area. Page Lake near Manitouwadge has been optioned, meaning someone has bought the rights to explore there, but nearly all of these sites are in areas not under claim where you can take time to look around.

The Schreiber/Winston Lake area is being prospected. The Winston Lake Mine has increased the interest in prospecting that area. In the Schreiber/Terrace Bay area many individual mines produced in the past. I think an amateur would want to prospect and poke around here. The book *A Trace of Gold* is about the Schreiber area. It is listed in the appendix.

Away from the Michigan Copper Range, copper mining has not been nearly as successful. North of Sault Ste. Marie, about an hour's drive, is the Mamainse Mine location. In the middle 1800s at least three copper mines opened here. None were successful. The area is now a park and a pretty drive. Another report tells of traces of gold found when ore from Michipicoten Island was assayed for silver and copper.

Beardmore-Geraldton

Beardmore-Geraldton is a region with a long history of gold production. Today exploration goes on in the district. Some geologists are looking in similar rock formations near the old gold mines. One company is actively drilling in the belief that a major fault similar to the gold-bearing one mined between the 1930s and the 1960s surfaces east of the old mines. Also, one advanced drilling project is using new theories to explore the Brookbank Prospect.

Near the towns of Beardmore and Geraldton the names of the now-closed mines include: the Hard Rock, the Magnet, Tombill, Jellicoe and at least six other major mining operations. The biggest of these mines produced astounding amounts of gold in its time.

One of the biggest mines in the Geraldton region was the MacLeod-Mosher mine. It produced well over one-and-a-half million ounces of gold between 1938 and 1970. Its silver production was over 107,000 ounces during the same period. At today's prices that would be almost three-quarters of a billion dollars.

Top: The original excavation of the Hard Rock Mine near Geraldton. Bottom: The Hard Rock Mine.

Certainly it would rank as an interesting area to check out if you are serious about finding a little loose gold dust.

The Hard Rock Gold Mine produced 269,081 ounces of gold and 9009 ounces of silver between 1938 and 1951. At that time the gold ore played out. While the names of the principal

MacLeod-Cockshutt shaft, Little Long Lac, 1936.

owners have merged a few times in the past, no one is letting go of that claim.

The Leitch Gold Mine produced 861,982 ounces of gold from 1936 to 1968. At the same time, 31,802 ounces of silver came from the mine. In 1980 the owners of record, the Teck Corporation, noted that slag from the mine still lay on the property. This is after they had screened the material for a number of years. By reading between the lines of the report I have to conclude that they were making money by screening and processing the stuff the first miners left behind. Shades of the Chinese bosses of the Old West.

One of the first mines of the region was Little Long Lac. Between 1934 and 1956, 605,449 ounces of gold were recovered along with 52,750 ounces of silver. Now that the gold has played out there, the search for more gold and base metals has intensified.

There are some excellent books on these towns and mines by regional authors. Two I found interesting are *Muskeg Tours* and *And the Geraldton Way*, both by Edgar Lavoie. For more complete bibliographic information check the back of this book.

The Ontario Ministry has an Historical Research Project promoting interest in the province by putting out information on old claims and occurrences. The concentration has been on the Beardmore-Geraldton area. Historical information on other areas of northwestern Ontario, much of which is not covered by this book, is included. The objectives of the Ministry are to document the old prospects, locate them and put the information in a form people can find and use. This information is in Thunder Bay at the Resident Geologist's office.

Thunder Bay

Near Thunder Bay a new surge of prospecting activity in the 1980s has occurred. Gold and copper have been found in the Shebandowan Greenstone Belt. Snodgrass Lake, west of Thunder Bay, had a good showing of visible gold at one prospect. Drilling goes on at Gold Creek a little further west. Even the name makes it worth a stop just to shake out a pan or two.

The most famous of all the silver mines was Silver Islet near Thunder Bay. The records of the mine are almost unbelievable to read. Much prospecting for silver was done in the area. Finally, silver was found in quantity in 1877 when drill core samples showed rich ore. The first major find was a pear-shaped deposit only 335 feet long. But it yielded two million ounces of silver! A second ore body, shaped like an upside down cone, ran for 285 feet. Descriptions of the mine tell of native silver being visible and twisted into hooks and spikes in the shafts.

Silver Islet is just that, a tiny island down at the end of a long peninsula southeast of Thunder Bay. Historical data, pictures and books are available about the mining of this tiny island. (Check the bibliography in the back of this book for titles.) Though just a very small island when the first shaft was started, the size of the island grew as the tailings from the mining were dumped into the bay to provide a breakwater. In time, buildings, a headframe, piers and landings were erected on the island. Most are now gone. You might want to drive down through Sibley Park southeast of Thunder Bay to find Silver Islet.

West of Thunder Bay, just off the highway leading to Atikokan, are the ghost towns of East Silver Mountain and West Silver Mountain, silver mining camps that worked part of the Quetico Fault described in Chapter Two. These mining camps produced quite a pile of silver in their heyday. Hundreds of

Double jacking, Silver Islet, 1880.

people lived and worked there. Today very little is left of the towns. Look for the historical marker road signs.

Near Thunder Bay, the Inco Shebandowan Mine is working a copper-nickel lode. Near Black Bay native copper has been found on the beaches. One prospector told me he finally found the copper vein there by climbing two hundred feet up the side of the cliff back of the beach.

The first known mining efforts on the Canadian side of Lake Superior were made by Colonel John Prince in 1846 or 1847. Copper and silver were found in a vein in the area now called Prince Bay, between the Pigeon River and Fort William. No records of production of copper exist but apparently some silver was found. A roadside marker shows the area where Prince prospected.

Atikokan

The gold areas west of Thunder Bay are broken into three sections: the Shabaqua-Shebandowan Lakes Area, the Atikokan Area and the Mine Centre Area. The gold and silver production of these three areas is small in comparison with Beardmore-Geraldton, but a total of 59,000 ounces of gold and 174,000 ounces of silver were found in all the mines. The deposits all fall in a curvy line running along the Quetico Fault. Highway 11 connects all the mining towns. The map showing mine locations puts most of the interesting ones very close to this highway. Each of the sub-areas appear to be where major geologic structures like greenstone batholiths come to the surface.

In the Atikokan area, interest still exists among prospecting firms and the Ministry of Mines. The area has been actively explored for minerals since the late 1800s. Its total gold production is listed at 8,000 ounces and total silver at 1,500 ounces, but most of this production was when prices were low. At today's prices this would amount to $3,500,000 (US). With that figure in mind, I could put up with some of the blackflies at the Blackfly Occurrence.

Mines in the Atikokan area were producing gold in the 1890s. The Harold Lake mine produced 1,131 ounces of gold, the Sawbill mine over a thousand ounces of gold, and the Hammond Reef and Elizabeth mines produced small amounts of gold. The Sapawe Gold Prospect was first discovered around the turn of the century. Mining did not start until the 1960s when diamond drills brought up cores showing gold. The Allegheny Mines Corporation now is attempting to rework the Atiko-Sapawe gold mine in the greenstone. In Minnesota, just south of this area a lot of gold exploration was done. I talk about these sites in the Minnesota chapter.

There are many abandoned mines near Atikokan. I have heard from fishermen that two abandoned gold mines are on the east shore of Sawbill Bay in Marmion Lake, accessible by boat or by snowmobile from an old logging road. At last report

Buildings of the Elizabeth Mine camp.

the road was accessible by four-wheel-drive vehicles. There was a tailings pile at this site, removed when the price of gold rose.

The Mine Centre Area was the site of the first of Ontario's gold rushes. The railroad opened up the area in the early 1880s, connecting Lake of the Woods to the rest of the country. Prospectors moved in to try their luck and were already working in this area when the word came from the south of the gold rush at Lake Vermilion, Minnesota. This rush and the gold found at the Little American Mine on Rainy Lake caused much excitement and spurred prospecting. This prospecting paid off for some; gold was found near present day Kenora. The Golden Star Mine produced 11,745 ounces of gold. The Foley Mine produced 2,043 ounces of gold.

Quetico Provincial Park has gold associated with it, too. The Old Moss Mine is just east of the park, and the Hackl Gold Occurrence is south of Middle Shebandowan Lake. That area is not too far by canoe from the Gunflint Trail of Minnesota.

At the far western edge of the range covered by this book is the area of the McKenzie Project. This is north and west of

Top: Moss Mine shaft, 1926. Bottom: Ardeen Gold Mine (Moss-Huronian Mines), 1934.

Fort Francis and International Falls. Here the Ontario Geological Society has found excellent gold grain counts in the glacial till. They are using a sampling method called "reverse core drilling," and they are panning. Similar work is being done in Minnesota south of this area. I describe the area and the theory in detail in the next chapter. A new discovery near Elizabeth Lake is being followed up in the Atikoka area.

The Ontario Ministry of Northern Development and Mines has excellent maps, books and reports detailing every gold mine, prospect and occurrence within the province. You will only be helped by the work their geologists have prepared. I recommend very highly that you look in the appendix for information on where to order or see the reports and maps that concern your locations of interest. You will not be disappointed.

Chapter 4
Gold Prospecting in Minnesota

Our power boat bobbed a few feet from the rocky shore. There at water level, almost hidden behind scrub birches, was the old Little American gold mine. "No," my guide said, "I've been by here many times. I've never gone in. I wonder what's in there?"

Minnesota has recently seen both a rise and fall of interest in gold prospecting. During the late 1980s much government land was leased for the exploration of gold, copper-nickel and platinum. The recession and lack of a significant commercial strike has seen a drop in the number of leases sold at the last few auctions, but many professional prospectors still work in the area.

There is an area of intense speculation near the Minnesota towns of Virginia and Gilbert on the Mesabi Iron Range. The drilling is in what is referred to as the Virginia Horn, a small area of intense folding and refolding of the quartz veins on the edge of a greenstone belt. Gold was found there when a trench was cut for a new railroad line. Three companies are drilling there, even in people's back yards in Gilbert.

East of Burntside Lake, near Ely, Minnesota, is the Raspberry gold prospect. At this site, gold was found in an outcropping in 1890. It has been worked off and on throughout this century. Test shafts and pits were dug and test holes were drilled, the last recorded drillings being in 1969 and 1972. Samples from this formation had visible gold. Another area near there is the west side of the southwest bay of Long Lake. Assays of ore samples from this site show gold. The west side of Shagawa Lake has some exposed granite that has veins of milky quartz and gold. At this writing, a gold exploration lease is being prepared for a mining company to work on Spaulding Bay in Shagawa. They apparently plan to set up on the ice and drill through the water. This will be interesting to follow over the next few years.

Drilling and sampling are occurring on the eastern end of Lake Vermilion. The work, being done near cliffs, can be seen from the gravel road. Watch for claim signs if you look there. There are also some old iron mine shafts sealed off in that area. You may want to ask permission to look around.

A new area of interest being explored in Minnesota lies west of the towns of Cook and Orr. Geologists have found gold in the gravel pits of the countryside. A Canadian company, Overburden Drilling Management Ltd., has developed a plan to trace gold grains in gravel back to the original vein. In effect, this is similar to watching the spread of smoke from a chimney. The theory is that if you find gold in the gravel and you can accurately determine the pattern of the glacier, your gold pan or drill core samples will lead you back to the original vein. The distance the gold has been moved from its vein is determined by the coarseness of the grains. The smoother the grain, the longer it was in the grip of glacial forces. (See figures.)

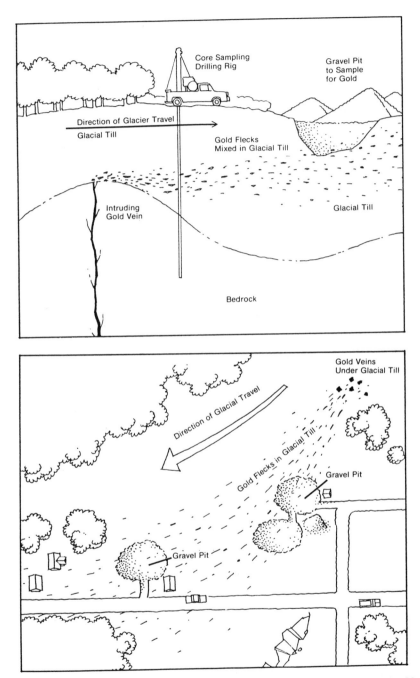

Figure 1 (top): Side view of gold drift in gravel. Figure 2 (bottom): Aerial view of gold drift, showing how gold followed the glacial path away from its original source.

The glacier scoured the bedrock, taking with it pieces of rock broken loose by the ice. Some of the rock in this region contained gold. When the ice melted, the rock and gold fell and was washed by the action of the retreating water. Don't think that this means the gold traveled hundreds of miles. Most of the samples show travel of up to only a few miles.

A lawyer from Duluth told me he had found gold and gold nuggets in northeastern Minnesota, but he would not divulge the location of the find. Mineral exploration companies think highly enough of this mention to have purchased leases in 1988 to drill in the entire region. You too, may want to spend some time panning for gold in some of the gravel pits found along back roads.

This underscores the amateur's advantage over the major mining operations; the price of gold has to be high enough to warrant the expense of extracting the gold from the surrounding rock. This is not a big problem for the amateur prospector or the hobbiest. If you pan a gravel shoal or rapids or a stratified gravel pit, much of the concentration of heavy metal has been done naturally for you. You will not need a large sum of gold to make your trip pay for itself.

Recent geologic analysis and mapping has shown the existence of greenstone structures similar to the gold-bearing rock of Ontario between International Falls and Voyageurs National Park. Geologists and prospectors will be combing the woods looking for outcroppings from these formations. You may want to look closely at some outcroppings, washouts and cliff faces for samples. Remember to respect park rules and boundaries.

Minnesota Prospecting in the Past

The following is a list of Minnesota counties showing the number of known gold occurrences in the Arrowhead Region: Carlton, nine; Cook, nine; Koochiching, twenty; Lake, two; Lake of the Woods, four; St. Louis, thirty-six. Some of these I will discuss because of their accessibility to hobbyists.

In Carlton County, gold traces were found about 1890 in test pits near the Silver Creek's mouth into the Kettle River and in railbeds of township 46N-20W, now known as Silver Township, south of Kettle River and west of Moose Lake. In this area, a number of test shafts were dug ten to thirty feet deep. Gillespie Brook, one mile or more to the north also had gold exploration done. Uranium was prospected in this area in the 1970s. No significant (or profitable) discovery was reported.

The quartz veins of the Thomson Formation near Jay Cooke State Park have excited interest over the years. The veins are mainly milk quartz with intrusions of other minerals. Watch that you do not get fooled by fool's gold from the pyrite intrusions. The best place to look in this area may be just below the Thomson Dam and near the Highway 39 bridge over the St. Louis River. At least five references to old gold test pits in this same area exist. Remember the rules about prospecting in park areas.

In Cook County, the Spalding Mine site on Lake Miranda is mainly a silver workings, but it has also shown some gold. Near the north/south portage on McFarland Lake a vein was shown to have silver and some gold. Another prospect showing gold, silver and other minerals is on Loon Lake. This prospect also showed arsenic among other minerals so be sure you trust your partners before you take them there! Gold Island on Saganaga Lake has been mined for gold and silver but, apparently, is not commercially viable. Much of this area is now in park boundaries, so do check the local rules before you make any holes.

In Koochiching County, old gold mines were operated on Rainy Lake and along the Minnesota/Ontario border country. These were hard rock shafts that bored directly into outcroppings that assayed gold. Because these mines were shafts in hard rock, there was no opportunity for panning or placer mining.

I have been in or near some of these mines; they still appear interesting. The Little American Mine, on Busheyhead Island in Rainy Lake, produced almost five thousand dollars worth of gold in the 1890s. This gold mine has a horizontal drift. I followed that drift to its conical end. On the way I had to cross what probably was the main shaft. It is filled with rock and water and appears to be safe. The adit is at lake level and is visible from a boat. This area is known as the Rainy Lake District, and the Ontario side has produced over a million dollars worth of gold over the years. Though it has a number of old mines and test holes—many named for the islands on which

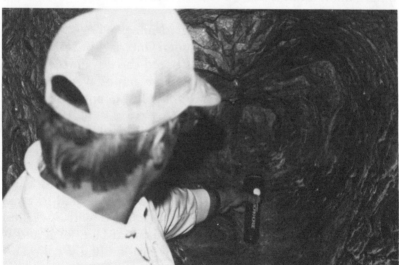

Top: Adit of the Little America Mine, Bushey Head Island, Rainy Lake, Minnesota.
Bottom: End of the drift at Little America.

they appear—remember that most of this area is now in the Voyageurs National Park.

Local legends in International Falls, Minnesota, say that, after the gold played out in the Rainy Lake mines, the ore tailings were brought to the Falls to be used for road beds. Legends say that more gold is under the tar streets than was ever taken out by the miners.

Also near International Falls, ask about or look for the Old Soldier Mine, the Big and Little American Mines and the Lyle Mine. In the Indus area, a number of positive tests for gold in the base rock have been reported, but these would be hard for the amateur to find. I would think that some of the gold may have washed out over the years and collected in low areas of older creeks.

St. Louis County, Minnesota's largest county, has the largest number of reported gold finds in the state. A sample taken from somewhere in Duluth at the turn of the century noted gold that could be seen with a microscope. This area was identified only as being on Lake Superior east of Minnesota Point and near Brewery Creek. It apparently was not enough to stop the Fitger's Brewing Company from taking the creek over for the brewery. This whole stretch of lakeshore was just torn up for the freeway extension. The bedrock was blasted to make tunnels and interchanges. The newly exposed rock would certainly warrant an interesting look. The removed rock is being used to create the shoreline of the new Lake Walk near downtown Duluth. So take your magnifying glass when you go on the Lake Walk.

Also in St. Louis County is the Lake Vermilion Formation. The gold rush of 1888 to Lake Vermilion was a major cause of opening up that part of Minnesota. A road was built from Duluth through the woods to Lake Vermilion, and prospectors came down from Ontario and up from the States. After a while, some people noticed that the iron content of the area was throwing compasses off. That led to the discovery of the huge iron deposits.

You can read the historical marker that stands where the main stamping mill once stood just off Highway 169 by Everett's Bay on Lake Vermilion. Surveyor notes of the 1920s showed deserted gold mines and an abandoned stamping mill just south of this area. The mines were scattered around the lake area, and the miners brought their ore to the mill to be processed.

The first town in the area was Winston City. In 1888 it had a population of over 300 people. The location is on the shore of Pike Bay between the Pike and West Two Rivers. Aronson's Marina is near the townsite and can be seen from Highway 169. For a few years there was a lot of activity with stamping mills, saloons and the other usual trappings of a mining town. But not enough gold was found in the region to support the town, and it faded away. Legend says that, when the nearby town of Tower was being built to house iron miners, all the steel needed for nails and such was taken from the old shovels and axes and other equipment left lying around at Winston City.

An outcrop on Ely Island proved to have a little bit of gold, but it was not economical for commercial mining. It does indicate that all looking is not in vain. The New York Mining Company had an operation set on the south shore. Nobles and Company had a location further to the northwest. A third operation was by the firm of Seymor and Company. N.H. Winchell, a geologist who worked through the area before the turn of the century, thought eight or ten companies worked in the area during the gold rush.

Gold Island on Lake Vermilion was named with hope that the mine shaft there would produce wealth. An old boiler was

Old boiler at the Theresa Gold Mine.

found at the site near the shaft, and it most likely provided the power needed to bring the ore to the surface. A stamping mill worked just north of this area on Trout Lake. The efforts to dig the gold included hauling these heavy machines by oxen up from St. Paul.

A gold mine was opened at Eagle Nest Lake, east of Soudan, in 1885. Apparently, not enough gold was found to make the venture worthwhile, but I don't know any more on this. It may be worth investigating.

A booklet entitled *Exploring St. Louis County's Historical Sites* by Charles Aguar lists some homey reports of gold prospecting. Some of his references are used in the report "A Compilation of Ore Mineral Occurrences, Drill Core and Test Pits in the State of Minnesota" by the Minnesota Department of Natural Resources. (Both of these are listed in the bibliography in the appendix of this book and are very interesting and useful to have along.)

One report is that near Echo Lake a mine shaft was sent down, equipment brought in and buildings built. Assays resulted in samples of measureable gold. Near Crane Lake were many spotty gold finds. Two reports of gold are named only by the owners at the time, a Mrs. Ole Hanson and a Mrs. Pachale reportedly discovered gold and silver on their farms. They opened mines, but I do not know the results. I don't know the value or accuracy of these claims, but it looks like enough gold was found to assay and interest hardy woods people in breaking rock.

Another mine was on the Vermilion River banks, not too far from Crane Lake. One shaft was deep enough to have two working levels. This mine was thought to be richer than many others, and, following a vein, the workers sank a shaft three hundred feet through volcanic rock. The mine was supposed to be shipping ore in the mid 1930s, but I did not find any record of gold shipments from this mine.

In 1907 a Mr. Albion Fenton of Minneapolis chipped off a piece of rock he thought to be iron ore from an outcropping. By the time he got around to having the sample assayed, he forgot where the outcropping had been! The site was not found again until 1924. A long time passed between refinding the ledge and opening up the mine, but the difficulty of getting into the area at the time and the lower price of gold could account for that. This site is six miles from Buyck, north of Orr.

Equipment ruins at the Theresa Gold Mine.

One rumor to pass on came from an older man who operated the wayside rest on U.S. Highway 53 south of Virginia, Minnesota. He stated that his buddies claimed their hunting shacks were sitting on gold. Their shacks were in the vicinity of St. Louis County roads 16 and 4.

Silver

The Minnesota DNR's book cited earlier for gold samples lists the following number of locations for silver in the Arrowhead counties: Carlton, two; Cook, eight; Koochiching, nineteen; Lake, one; Lake of the Woods, nine; St. Louis, twenty-one. Refer to the DNR book for locations of these finds. Especially interesting is that in St. Louis County most of the finds were from outcroppings and test pits. This indicates that they can be found by the amateur prospector.

Point of Rocks, near downtown Duluth, Minnesota, the site of a silver mine in the 1850s. Blasting removed much rock. Rumor says the best silver was still underground when the mine closed.

A silver mine was located in downtown Duluth in the 1850s in the area known as Point of Rocks. The mine was started with funds from investors and apparently turned up some very good silver ore. Unfortunately, a panic hit the market after the ore had been brought to the surface. The mine operators were unable to get together enough additional money to send the ore to be refined. After three years in a warehouse, the silver-bearing ore had tarnished so much that it was worthless. The ore was thrown into the lake as fill for the area now housing the Duluth Entertainment and Convention Center and the investors were out their money. No sign remains of the mine. Point of Rocks is being blasted now for road construction, so something may yet come to the surface.

Copper

Native copper has also been found in Minnesota. Specimens taken from the iron mine at Tower-Soudan are on display in

Chisholm at the Ironworld Museum. Ancient copper diggings are found along the French River northeast of Duluth. A roadside historical marker points to the place. It tells that "rumors of nearby copper deposits resulted in widespread prospecting." A town, to be called Clifton, was platted but never came to be much. The French River Mining Company, in 1864-1866, dug several test shafts but did not find commercially viable deposits. Test drilling is now being done in northeastern Minnesota for copper-nickel deposits.

Diamonds

Some companies are searching for kimberlite deposits in northern Minnesota. Kimberlite and diamonds are discussed in Chapter Six. One company drilling in the area South of Lake Vermilion has a lease for a lot of land but appears to be concentrating its efforts in one small area. I have heard that another company is drilling in Aitkin County. I have not seen information to substantiate this.

Northeastern Minnesota is a large area with very few people per square mile. It is quite possible that a major gold find is still waiting to be found. There are many trails, wilderness areas, gravel pits and backroads to explore. So take your rock hammer and gold pan along.

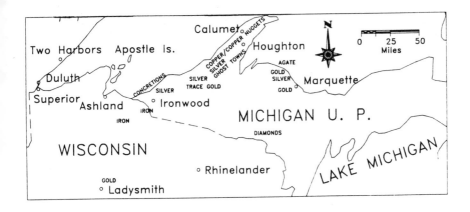

Chapter 5

Prospecting in Michigan and Wisconsin

There was some recent bulldozer work that caught my eye. We turned around on the highway, headed back and turned in. A short drive up a bumpy road led to barbed wire. Minutes later we were looking into the settling remains of yet another copper mine in Michigan's Upper Peninsula.

Gold

Fantastic reports of gold on the surface were a fairly common experience for Michigan gold prospectors and miners. Three major mines were opened at sites where the gold was assaying out at (reportedly) between $400 and $100,000 per ton with the price of gold around $25 (US) an ounce. Unfortunately, the gold played out just a little ways under the surface in all the mines.

The history of gold in Michigan goes back to the early days of the French voyageurs hearing reports from natives. Gold and copper deposits were major reasons for the United States government acquiring Michigan, but tragedy befell the discoverer of Michigan's first significant gold find, State Geologist Dr. Douglass Houghton. He drowned in Lake Superior before divulging the location where, according to accounts of the story, he had panned out enough gold to fill an eagle's quill. The city of Houghton still bears his name.

Although gold was one of the major reasons for early exploration in northern Michigan, today there is only one working gold mine in the Upper Penninsula, the Ropes Gold Mine, near Marquette, Michigan. The Ropes Mine is owned by Callahan Mining Company, and they worked it during the 1980s. At this writing, production has been suspended at the mine.

A lot of speculation and prospecting is going on near Marquette. The work is concentrated on the Marquette greenstone formation and in the Marquette Trough. Callahan Mining and other companies are still drilling, mapping and assaying rock samples from the Trough. A close look at a topographic map of the area gives a good idea of this formation's boundaries, west of Lake Superior past Marquette and south of the hills of the North Shore.

The Ropes Mine is working the Marquette greenstone formation. The success of the Ropes in the 1890s encouraged many other explorations in the Marquette County area. The Ropes Gold Mine was named after Mr. Julius Ropes, a chemist with many other interests including prospecting, who found gold in many rock samples and finally found the quartz vein motherlode when he chipped out samples from a moss-covered outcropping.

The Ropes Mine produced more than $600,000 (US) until the main ore body was exhausted about five hundred feet down. There is much speculation that with better mining and milling procedures more gold could have been produced. If you are going into the area, a review of the mine's history is interesting. Purchase "Gold in Michigan—Open File Report 80-1." It is listed in the bibliography in this book's appendix.

At least four other mines were sunk in the vicinity of the Ropes: the Michigan, the Gold Lake, the Superior and the Peninsula mines. The Michigan Mine was the only one to achieve significant production. Its total output was about $18,000. Some notes about the Michigan Mine come from Dr. A.E. Seaman, who has the museum in Houghton named after him. You may come across some covered shafts, trenches or pits as you prospect the area. Remember the rules about private property.

I have been told that open land policies exist on most of the land owned by the big paper and timber companies. This means they will not mind if you go through their woods as long as you do not damage the woods or leave litter. Much open policy land exists for prospecting in the Upper Peninsula. Just check first and save yourself hassles.

Panning has shown gold in many streams in the Upper Peninsula. Books from the state geological survey show unverified reports from more than twenty locations. These include the Flat River in Ontonagon County, the Ishpeming in Marquette County and the Little Sable River in Manistee County. There was also a report of a large nugget from the Victoria Copper Mine.

Panning is done every summer on the Yellow Dog Plains by amateur prospectors. The Dead River Basin also has shown small amounts of gold but not enough potential to bring in a big company. But you should bring your pan. One geologist told me that a college buddy worked his way through school by panning for gold in this region every summer.

Placer gold has been found in the Dead River, the Yellow Dog River and the Yellow Dog Plains area northeast of Ishpeming. Placer mining is when large hoses are used to spray water onto hillsides to wash gold (and everything else) down to a separating plant. It is very damaging to the environment because of the erosion problems it causes.

In Wisconsin a new mine has opened near Ladysmith. Sulfide ores are being brought to the surface at this site. Gold is a secondary product being milled from the ore. I would guess silver is also being found there. This mine is the first new mine opened in the United States portion of Lake Superior in recent times.

Gold and silver exploration has had minimal success in the part of Wisconsin covered by this guide. There was some exploration and mining farther south. Gold Diggers Creek near Eagle River in Florence County was the site of some prospecting and mine digging, but no significant gold was found at this location. Some visible gold flakes have been found in Dane, Crawford and Marathon counties. But, as one geologist put it, you can find gold anywhere if you want to sift enough earth.

Gold is known to be found as a by-product in many non-gold mines throughout the Lake Superior region. Samples of gold and gold-bearing ore have been brought to the surface from the copper mines of Michigan, test holes in Minnesota, and silver mines of Ontario. These samples always excite interest but have usually proven uneconomical to open full-scale operations. Excellent samples listing the location of their discovery are on display at the A.E. Seaman Museum in Houghton, Michigan, as well as at some local historical museums.

The following are some of the samples of gold and gold-mixed other minerals that have been found, as seen at the Seaman Museum. Gold and gold-mixed with molybdenite has been found in the Michigan Mine, Ishpeming, Ontonogan County, Michigan; a gold sample was found at Kirkland Lake, Michigan; native gold in quartz came from the Hollinger Mine in Porcupine, Ontario; a gold sample taken from the Parkhill Mine, Ontario; gold was found at the Dome Mine, Porcupine District, Ontario; and a serpentine (a green crystaline rock) sample came from the Ropes Gold Mine. Time taken to view these specimens at the Seaman Museum can give a good idea of what specimens look like. The Museum is on the campus of Michigan Tech.

Silver

Houghton County had significant silver production in the early part of this century. The Isle Royale Mine produced thousands of dollars worth of silver from 1909 until 1920. The North Kearsarge Mine produced silver from 1889 to 1892. Other silver-producing mines included the Franklin, the Osceola, and the Quincy.

The following Michigan counties also list silver finds: Baraga, Dickinson, Iron, Marquette, Ontonagon and Keweenaw. Some of the silver found in Keweenaw County are described as being little bells hung on copper wire. The Rockland Mine in Ontonagon County has a story about two kegs of silver nuggets being lost and found in an abandoned shed years later!

No working silver-only mines are found today in the Lake Superior region. The low price for silver (at this writing less than nine dollars [US] an ounce) does not justify investing all the equipment needed for large scale mining. Many mines do produce silver because it is in the rock they are processing. For example, the White Pine Mine of the Copper Range Company in White Pine, Michigan, is mainly interested in mining copper, but is currently producing silver because the silver-bearing ore is of sufficient quantity to make refining commercially worthwhile.

Silver mining was attempted from 1910 to 1915 on the edge of Lake Superior along the Iron River in Michigan on an outcropping of the same ore body currently being mined by the White Pine Mine. In this area many test holes and some shafts were dug. Mines there never panned out probably because the early miners did not understand the geology of the area. The silver there is not in veins but mixed into the surrounding copper-bearing rock.

The prospectors worked on the assumption that the silver would be in veins like the veins at the edge of the White Pine Lode. That vein is apparently the result of the heating processes of the early tectonic activity forcing the silver out the sides of the hot volcanic rock. It did not prove profitable to mine at the time.

Samples at the Seaman Museum in Houghton showed the following interesting samples: in Michigan, silver from the Michigan Mine near Ontonagon; samples of silver from the Kearsarge Vein and Kearsarge Mine; silver and wire silver on copper at the Osceola Mine; leaf silver found by the Michigan Mining Co., Ontonagon; silver crystals on copper and leaf silver from the Wolverine Mine; a sample of crystalized silver from the Franklin Mine has the appearance of a dacshundt; silver and copper sample from the Adventure Mine in Greenland; silver from the Mohawk Mine in Keewauna County; silver crystals on copper from the LaSalle Mine. There were samples of silver from Ontario, the Silverfield Mine and King Edward Mine in Cobalt, the Mann Mine in Goganda, and silver in argentite from Silver Islet. Silver samples from the North Shore of Minnesota were also at the Seaman Museum.

Copper

In the 1840s, a Michigan copper rush was started by the findings of the state geologist, Douglas Houghton. The rush began petering out until, according to the story, a man named Samuel Knapp needed shelter. He was digging out a cave to make room when he noticed that the material he was digging was not natural fill. For the next few days he kept digging out the pit. At the bottom he found a piece of copper weighing over six tons propped up on old oak timbers. He then realized that someone before him knew how to find copper.

An 1862 Smithsonian Institution report listed pits near the end of the point, in the Rockland, Quincy, and Pewabic areas and on Isle Royale. Noted pits near Fulton and Calumet also existed. Almost every mine opened was in the location of the ancient pits.

Owners of working mines are aware of prospectors straying onto their properties. You would be better advised to look in the vicinity of closed workings when prospecting. In the Upper Peninsula the old smokestacks and foundations of abandoned mining activities can be seen along the whole Copper Range from the reconstructed cabins of Victoria to the fort at Copper Harbor. I have talked with local folks who pointed out interesting areas where I could poke around.

Cabins in various stages of reconstruction at Victoria, Michigan, site of early copper mines.

Abandoned mine tipple near Hancock, Michigan.

The underground miners are also very concerned for the safety of amateur prospectors who may fall into an old mine shaft. Their concern is very real and not just based on a love of amateurs. In the event of a calamity, it is the miners who will have to climb down into the shaft to rescue or recover. They tell me they do not want that chance for glory. So be careful in the vicinity of mine adits.

Old copper workings can provide a great day of activities. In Houghton/Hancock, Michigan, there is a museum to the underground workings of the copper mines. One shaft of the Homestake Mine there went down over ten thousand feet. Along the Copper Range are old tailing piles and the ruins of buildings and towns. You can find a lot of help to start your search from tourist information centers. Remember to use the same detecting skills needed to find gold when you are looking for copper.

Iron

While iron does not have the same lure as gold to prospectors, the Iron Ranges of the area are interesting to explore. Near Hurley, Wisconsin, and Ironwood, Michigan, are the remains of the Gogebic Range. Hundreds of underground mines, working railroads and towns were there in the mining heyday. There is still a tipple standing in one community and other mining evidence around the region, mostly shaft mines as opposed to the open pit style of mining found in northern Minnesota. There are lots of interesting variations of iron to find. I describe some of these in the last chapter of this book. It may well be worth an afternoon to look around this area.

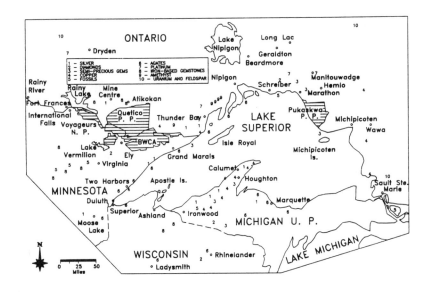

Chapter 6

Silver, Diamonds, and Semi-precious Gems, Copper, Ghost Towns, Petrified Wood, Fossils, Concretions and Odd Rocks

The open pit mine was sunny and dusty as we looked for shark's teeth in the discarded overburden. My three-year-old daughter, in overalls with hair in pigtails and face smudged by red dust, sat cracking rock on rock. She did not want to leave until she found a fossil.

While prospecting for gold, you may come across various minerals and fossils listed here. I add this information about them just to alert you to keep your eyes open. Other minerals and gems may be found locally. So ask around in areas where you prospect.

Silver is brought up at many copper and nickel mines today, and there has been silver mining in the past. Active diamond prospecting occurs today. In the rock and the glacial till of the Lake Superior region you may find diamonds and other gems. Gems are tricky for amateurs to find because they are small and do not appear in veins, as gold or silver may. There are agates, minerals and fossils, too. Take time for research if you are at a museum. Take the time to look if you are in a known location. Finding just one diamond should make you very proud of your prospecting.

Silver

Silver is a precious element found throughout the region both in pure veins and blended with other rock. Historically, some very profitable silver mines, especially in Ontario, have been opened. Silver is often found along with gold, nickel or copper. It is mixed with the same sulfide ores as these minerals. Since the ore is being brought out of the ground, mine operators might as well take the silver from it.

Silver is rarely found as the bright, shiny metal your silverware is made from. It is usually black or dark brownish in color. It would be much the same color as tarnished silverware. This tarnished appearance makes it look like a useless or unwanted ore sample. A crack of a sample with a rock hammer may yield a view of the shiny silver metal.

You may use a gold pan to look for silver. The action of panning is the same but the silver behaves a little differently. Its density is less than that of gold so it does not sink as rapidly. Care must be taken when panning for silver to wash with a more gentle action. This will help prevent losing the silver grains. Silver is not necessarily found with the black sand that accompanies gold. But I find the black sand just about everywhere I pan.

Rock-bearing silver is often found shot through with veins and tiny fingerlings of silver. Removing the silver requires a crushing and milling process much like gold. That should not stop you from looking for ore samples, however. If you really want to know what you have found, take the sample to an assaying firm. A nice chunk of silver ore would be a real fine trophy of an afternoon spent in the hills of the North Country.

Diamonds

Diamonds of gem and industrial grade have been found in the glacial till of central Wisconsin. There is a strong rumor that a subsidiary of the DeBeers Diamond conglomerate has been finding pieces of diamond in the glacial till near Kirkland Lake, northeast of Beardmore-Geraldton, Ontario. Rough diamond looks like clear quartz and will scratch quartz. Today, the prospecting being done is to find the place in the glacial till where the diamonds originally came.

The prospectors and mineral companies have been searching for kimberlite deposits, the host rock for diamonds. Kimberlite is an ultra-basic rock which occurs naturally in deep, cylindrical deposits called pipes. No satisfactory explanation has been made for their creation. A diamond is a carbon-based mineral formed under extreme pressure and heat. It is also known that the kimberlite deposits are fairly recent intrusions into the rock of the region. For diamonds to form, carbon was introduced into the kimberlite while the kimberlite was being formed.

Kimberlite is the term for both the rock in which diamonds are found and also the formation itself. Some theories look at kimberlites being formed in the vents of cooling volcanos. Other theories look at lava intrusions into fractures in the earth crust.

Kimberlite pipes are almost universally found in groups, like herds of elephants. If there is one kimberlite pipe found, others should be nearby. The simple existence of a kimberlite deposit does not mean the presence of diamonds. Only one in ten kimberlite pipes will have any diamonds. Only one in one hundred pipes will be commercially viable. Because there have been no commercially viable kimberlite pipes found thus far, the financing for diamond exploration is not readily available.

The search has been centered in the Lake Ellen area just south of Crystal Falls in Iron County, Michigan. Plotting of glacial activity has led experts to search this area. Kimberlite deposits also have been found south of James Bay in Ontario, in the mines of Upper Canada in Ontario, in Iron County, Michigan, and near Tower, Minnesota.

Garnets and Other Semi-Precious Gems

Many other semi-precious stones can be found loose in the till or embedded in rock. Finding these rocks is a matter of patience and skill in recognition. If possible, carry an identification book or some pictures if you intend to look for semi-precious gems, gems that have a dollar value to a buyer. Samples of local gems can be seen at county museums, rock swaps and historical places. If no rock samples are easily visible, ask the museum curator if any samples are stored away that you may view. Take time to look at the samples closely. This will aid you in recognizing them in the field.

Although garnets are found in Michigan, almost none are of high-gem quality. Two areas in Ontario that have yielded garnet are near Schreiber and north of Thunder Bay towards Armstrong. Some stones that appear to be garnets are actually pseudomorphs. This means that the garnet mineral has been substituted by another mineral but the stone still retains the crystal shape of garnet. You can see specimens of pseudomorphs at the Seaman Museum in Houghton.

Agate is the most famous stone found in the region. Agates are always a draw to stone hunters, and you can find small agates

Samples of Lake Superior agates.

on almost any beach or in any gravel pit in the region. Agate is often broken into fragmented pieces which, when wetted, readily show the bandings.

Some large agates do exist. The biggest known agate is 108 pounds. It was found near Moose Lake, Minnesota, and is displayed at a bank there. Some claim that the specimen is not a real agate, but that may just be sour grapes. Many agates are found in the one- to three-pound range with some whole agates going up to five pounds.

Some rare agates to look for are the purple-banded agates found at Seven Mile Point Beach and the black and white onyx agate found near Black Creek, both in Michigan. If you are not familiar with agates, look in local jewelry shops, museums or tourist stores. Lake Superior agate is the Minnesota state stone. A good book on the subject is listed in the bibliography.

Amethyst has always been my favorite stone from the region. It is a quartz crystal that has been colored purple by iron influences. The amethyst crystals in this region grew only in veins and vugs of a fault line running through the granite north of Thunder Bay. It is the official mineral of Ontario, and people of Thunder Bay seem especially proud of it.

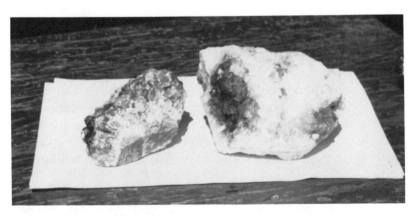

Large amethyst crystal formations.

Only four mines exist where you can find amethyst. These are all privately owned and allow rock picking for a fee, and are located about fifty miles from Thunder Bay. If you are in Thunder Bay, look in at the stores that are operated by the mine owners. You may purchase samples or finished stones there.

While you browse, ask to see the grinding and polishing machines at work. It is fascinating and amazing how much time is needed to bring the beauty out of a semi-precious stone. Some samples I have seen included amythest with flourospar that had been found in the Port Arthur, Ontario, area and amythest and quartz from Silver Mountain, Ontario.

Tiger's eye, a stone with parallel lines that reflect when polished, can be found in some of the iron-rich areas. Binghamite is a gem stone like tiger eye with silicified fibers of goethite, and is found near Crosby, Minnesota.

Chatoyant is an opaque gemstone with a hardness of seven. It is found in secondary deposits and vertical cracks in the sedimentary shale formations of the Minnesota Iron Range. Unlike tiger's eye or binghamite, the fibers in this gemstone are seldom straight or parallel to each other, but are usually masses of bent, crumpled and stitched strata.

Petosky stone, the Michigan state stone, is the fossil remains of colonies of coral reefs that grew near Petosky, Michigan. The coral has since been turned to calcite. Found on the beaches of Charlevoix County, Petroskyite is a fossil that appears to be an agate. It is not quartz but is listed as a calcite in mineral books.

Copper

Copper mines, holes, test pits and slag piles can be found all along the Copper Range of the Keweenaw Peninsula. Isle Royale and the North Shore of Minnesota above Duluth also have evidence of copper workings. Along the Ontario shore were some prehistoric diggings. These are fun to explore and prospect. The raw copper nuggets that are found make nice jewelry or paperweights and may aid you when telling your story.

The history of the copper mining in the region goes back thousands of years, long before current Indian history could even be remembered. Some people claim that the copper found in the great pyramids of Egypt came from the Lake Superior region. They point to the time 3,500 years ago when the Great Lakes were in the Nipissing Stage. During this stage three of the lakes were at 605 feet above sea level. There were open rivers to the sea both via the St. Lawrence and south from Chicago to the

Mississippi. The thought is that open boats could have made the trip to the ocean without having to portage great weights of copper.

There were over 5,000 prehistoric copper workings around the lake. Most were in the Keweenaw. These pits are still a great mystery to historians. Carbon tests show the pits to be 3,500 to 4,000 years old. There is almost no evidence of the people who worked these pits—no bones, no utensils, no pottery. While evidence exists in other parts of the region of the Laurel Indians from the same time period, no one knows much about the workers of these pits. Suggestions have included Vikings, Russians, Winnebago Indians and Central American peoples. One Indian legend describes a fight to drive "white people" from the area. Look to the local museums for information and relics such as stone hammers from that time.

The Voyageurs who first traveled in the area brought back reports of fantastic copper and gold mines in the region. They had found one great slab of copper that had weathered free. That particular piece was well known to the Native Americans. It had provided copper to them for centuries before the coming of white explorers. That three-ton piece of pure copper from

This sign marks the site of the first white man's attempt to mine copper near Victoria, Michigan.

the Upper Peninsula is in the Smithsonian Museum in Washington.

The first effort by whites to mine the region for copper was made by Alexander Henry in 1771. He led a group of men to mine near the town of Victoria, Michigan, while working for a company started by the king of England and some nobles. They did not do very well in their efforts, and the weather finally drove them out. You can find the location of that first mine by the reconstructed town of Victoria, Michigan, near Ontonagon. The cabins are just off a paved road that swoops through valleys and up hills. A little way past the townsite is a marker showing the original site of Henry's digging.

Raw copper pieces can be found in many parts of the region. Loose pieces of copper that have naturally weathered free are of most interest to the average enthusiast. Copper can be found along some beaches, near river gorges and in terminal moraines. You will also want to look in abandoned mine tailings. A metal detector can be helpful in this effort.

Float copper is the term that describes loose pieces of copper that have been weathered or torn loose by glacier action. When the glaciers last moved across the Upper Peninsula, much copper was left behind in the glacial debris. Some of these pieces are very small, weighing less than an ounce. Some are very large with a few noteworthy pieces in the region weighing over five thousand pounds. Needless to say, the term float copper does not refer to floating on water.

Pieces of copper are variously described according to their shapes. Sheet copper is thin like a tree leaf or foil. It has been found in sheets as large as the seven foot specimen at the Seaman Museum. Copper "skulls" are also on display at the Seaman Museum. These are not human skulls but they do have the same shape and size as my son's head. I found them a little eerie.

Copper is generally in combination with other mineral or rock compounds. Some interesting specimens of these copper combinations reside at local museums. At the Seaman Museum the minerals are displayed along with the specific locations of discovery. Some of these interesting specimens of copper combinations are: crystalized copper cubes from the Copper Falls Mine in Michigan; the largest specimen of sheet copper in the world and those copper skulls, specimens from the Tower-Soudan Iron Mine in Minnesota; a copper oxide sample taken

Large piece of float copper on display at the Copper Range Mining Company.

from the Isle Royale Mine in Michigan; samples of very beautiful copper and quartz crystals from the Minnesota Mine in Michigan and crytalized copper from the Central Mine in Michigan. Just seeing these samples can stir interest in finding your own.

Copper in mineral combinations is what concerns most of the large scale mining operations. Copper-bearing rock is not all that interesting to look at, however. I think pieces of pure copper or some copper in crystal are great for showing off.

Isle Royale was the site of some prehistoric and recent copper mining activity. The Indian people that mined the region before the whites appeared found some copper on the island. Early historical tracts refer to Isle Royale with the native name "Minong." Accounts from the 1800s tell of stone hammers by the thousands lying in and near some pits. These hammers needed to be carried from the mainland in a canoe. It is deduced from counting the stone hammers and the number of pits that possibly thousands of Indians worked the area for over a period of one to two thousand years. Isle Royale is now a National Park. The remains of mining adventures of both Indians and whites is preserved and may be viewed in the wild. Remember the rules about artifacts and just look.

Large pieces of float copper have been found over the years by wanderers and prospectors. Not all the big pieces necessarily have been found. In October 1987, two hunters found a piece of copper protruding from the ground. They tried to dig it up but could not do so. The piece turned out to be on Copper Range Mining Company property so the mine was contacted. The mine gave the hunters a finder's fee for that 5,950-pound piece of pure copper. That piece is, at this writing, on display at the front of the administration building at the mine headquarters. A 147-pound chunk of pure copper was found near the Mamainse Mine area in Ontario when Highway 17 was being built. Other large pieces may still be found.

When you are looking for copper in the tailings of old mines, I would recommend two things: use a metal detector and watch for crystals. The metal detector can tell you quickly that an area

Abandoned equipment at the Hill-Annex Mine near Pengilly, Michigan.

has already been fairly well-cleaned out. It may also pinpoint that wonderful specimen that lies just below the dust. Be sure to set the levels to screen out some of the background readings that may come from microscopic copper.

The crystals to watch for grew in vugs and vesicles which formed from gas bubbles in the cooling lavas. These hollow spaces left room for crystals to grow. The crystals were fed by water loaded with minerals that percolated through the hollows. Miners who worked the Gogebic Range told me of looking into basketball-sized pockets of beautiful crystals down in the mines. But they were not allowed time to collect samples. They had to load the crystals along with the ore and ship them to the surface. So you may find some of these crystals now in the tailings piles. Also look for samples of rope lava and other interesting specimens brought to the surface and discarded.

Fossils and Petrified Wood

Shark's teeth that are millions of years old rest in northern Minnesota. Fossils of shark teeth have been found in cretaceous

Fossils and shark's tooth found while prospecting.

ore removed as overburden at some of the open pit iron mines. Other fossils, including trilobites, have been found in the Hill-Annex and Douglas Mines in Minnesota. Various fossils can be found from the Cambrian and Ordivocian periods, approximately 450 million years ago. My three-year-old daughter even found fossilized worm tunnels in the overburden. Fossils of *Phacops rana* have been found in northern Michigan.

Many fine samples of fossils may be found on St. Joseph's Island, Ontario, on the east side of the lake and accessible only by boat.

Petrified wood can be found in gravel pits and mining piles. The wood pieces I have found are small. It is easy to see the old wood grain and cells in the pieces. I have not seen pieces big enough to identify species. I would guess others have found some. Fossilized wood has been found for certain in the overburden from the Douglas mine and the Hill-Annex Mine.

Samples of pahoehoe lava have been found in some of the mines of the Upper Peninsula. Pahoehoe lava has the appearance of soft mud that has rolled and solidified. It has no intrinsic value but is an interesting find. It is evidence of past volcanic

Some fossils.

activity that was later covered by other geologic occurrences. One very nice sample was found at the twenty-eighth level of the Isle Royale mine in Houghton County, Michigan.

Concretions can be found on beaches near Hurley and Ironwood. These are sandstone-like marbles created by wave action on fine sand. Very large concretions are found inland near Thunder Bay. Some are shaped like the Buddha, and others are just like playground marbles. These have no intrinsic value but certainly are fun to have as a remembrance of your trip.

Minerals

There are many special minerals found in the region which can be polished to look like gems. I list some so that those with a careful eye will know to look for more than just agates. If you have an interest in minerals or want samples for a collection, keep an eye out for rock shops along the way. They often carry samples that came from just outside their back doors. You can buy their samples or ask them where to go find your own.

- Feldspars and uranium are found in Ontario.
- Manganese ore has been found near Chisholm, Minnesota.
- Nickel and other minerals are mined or found throughout the whole region.
- Iron from the region can take on some beautiful characteristics. Specular hematite is a high-grade iron mineral that is often found in iron mine tailings. It is very reflective, almost appearing to be silver when held to light. I have found many pieces of specular hematite and saw one highly polished sample from the Beacon Mine in Marquette County, Michigan.
- Iron ore breccia, a pretty combination of purples, yellows and greens, can be found in the dumps of iron mines. A nice sample was from the Gary Mine near Hurley, Wisconsin.
- Goethite, often called limonite, is a tumbling grade gem sometimes found on the Iron Range. Its iron content varies from 54% to 61%. It is very similar to hematite. Limonite can contain dendrites or false ferns. This is the result of precipitation of limonite into taconite. It can be very pretty.
- Sparkling galena from the Dead River District of Michigan makes a pretty find. These iron combinations can be easier to find using a metal detector.

- Barite, from the Montreal Mine in Wisconsin, is a crystallized, whitish, gray and brown rock found in layers.
- Jaspilite is a tumbling grade gem similar to jasper. It can be found in the vicinity of Jasper Peak near Tower, Minnesota.
- Saponite, a quartz-like rock, is a silicate of magnesium aluminum that has been found in the Houghton, Michigan, area in the Laurium Mine.
- Prehenite, a green, jade-like rock that formed from dripping water has been found in Osceola, Michigan.
- Analcite, a crystal that is very pretty, has been found in the Phoenix, Megaunee and Isle Royale mines of Michigan.
- Thomsonite, a cross-shaped crystal, can be found on Thomsonite Beach along Minnesota's shore line.
- Ochres, both red and yellow, are found in the area around Lake Vermilion, Minnesota, especially by some of the prehistoric sites near Daisy Bay and near Tower. Ochre has no gem value but is used in dye and paint making.

Ghost Towns

Throughout the Lake Superior region are many other interesting and fascinating things to catch your attention while puttering through. Near dusk on a late summer eve in the Upper Peninsula, Dad and I stumbled onto, and almost into, an old iron mining glory hole. A great issuing of black night came from that hole protected by only a few strands of rusted barbed wire. It was compelling to look into even though a slip on moist rock could mean a sliding fall hundreds of feet to a very hard end. The magic and danger of finding that glory hole still come to me today.

There are ghost towns around Lake Superior as real (or unreal) as any of the Old West. Some of them are abandoned mining camps with no living residents. Others were logging camps back in the woods. Some were watering stops for the trains that crossed the wilderness. I have marked some of these townsites on the chapter maps. More information is available from local people and historical societies. I have listed some books in the bibliography that tell of ghost towns in Ontario and Minnesota. Please remember that someone does own the property, and do not deface or destroy buildings or equipment.

Edgelake Gold Mine, Toshota Lake.

Old mine trolley, Edgelake Gold Mine, Tashota Lake.

I have seen old towns, buildings, and homesites all around the region. On the Iron and Copper Ranges were many small towns that are now gone. Some were swallowed up by the mines while others disappeared with the closing of railroad lines. A friend and I found an old map of the land where his summer cabin is located. Within two miles was a town shown which today is not even remembered by any sign, marker or road to the location. It was well worth a hike to check out that old town. While hiking Isle Royale, I encountered the remains of old mining and hotel buildings. These are now protected within the National Park. If historical sites interest you, do a little research and you can discover them for yourself.

Go out and look for your own treasure. My fingers start to itch even as I sit and write. They want to be out picking stones and swirling the gold pan. Soon it will be spring, and the prospectors will be out ahead of me. Someone will find a big gold strike soon, probably not a geologist from a multi-national company. It most likely will be an independent prospector working a claim on some knowledge and a hunch. Perhaps you will have a hand in that next gold strike. Get out and look, enjoy the beautiful Lake Superior region, clean up after yourself and good luck. One of us will find it.

Sources of Information

Sources: Books and Pamphlets

These books, pamphlets, and reports are available at modest cost. Ask for current prices when ordering.

Report of Activities 1990 Resident Geologists, Ontario Geological Survey Miscellaneous Paper 152, edited by K.G. Fenwick, J.W. Newsome and A.E. Pitts Queen's Printer for Canada, 1991. Available from Public Information Centre, Ministry of Natural Resources, Room 1640, Whitney Block, Queen's Park, Toronto, Ontario M7A 1W3 or telephone 800-668-9938. (Check price and shipping.) Absolutely excellent reference to take along in Ontario. Clear maps, good arrangement of locations and more information than you could possibly use. (Does get updated.)

Rocks and Minerals Information Sources. Compiled by the Ministry of Northern Development and Mines. Printed in Ontario by Queen's Printer for Ontario. Order from Public Information Center, Ministry of Northern Development and Mines, 99 Wellesley St. W., Room 1640, Toronto, Ontario, M7A 1W3. Excellent catalog of available maps, reports and surveys, prices listed—English and French. Also, lists of bookstores, specimens and identification tables, names of mineral and lapidary dealers, geoscience clubs, mineral and fossil exhibits and periodicals. A must for anyone who is looking in Ontario.

Report 231: A Compilation of Ore Mineral Occurrences, Drill Core, and Testpits in the State of Minnesota 1985 (May be updated), Minnesota Department of Natural Resources, Division of Minerals by Dennis P. Martin, Hibbing, Minnesota. A comprehensive listing of all known sources of all minerals in the state. Heavy emphasis is on St. Louis and Cook counties in NE Minnesota. An accurate guide with specific locations and findings listed.

The Geology of Gold in Ontario, Ontario Geological Survey Miscellaneous Paper 110. Edited by A.C. Colvine, 1983. Some general information and many specific articles and maps pertaining to the entire province. Does get technical in many places.

Silver Islet, Striking it Rich in Lake Superior. Elinor Barr Natural Heritage/ Natural History, Inc. Toronto. 1988. Excellent data, information and pictures about the mining history of Silver Islet.

Confessions of a Cornish Miner; Silver Islet 1870-1884. James Strathbogey. Porphry Press, Thunder Bay, Ontario, 1987. Personalized account of life on Skull Island, later called Silver Islet.

". . . and the Geraldton way." Edgar J. Lavoie. Corporation of the Town of Geraldton. 1987. A history of the Geraldton gold mining region from early years through the gold period.

Muskeg Tours, Historic sites of the Little Long Lac Gold Camp. Edgar J. Lavoie. Squatchberry Press, Geraldton, Ontario. 1987. Pictures and text guiding the reader to sites of historical interest in the gold camp area.

Ghost Towns of Ontario, Vols. I and II. Ron Brown. Cannon Books, Toronto, Ontario, 1983. Pictures and stories of fading and ghost towns.

Property Visits and Reports of the Atikokan Economic Geologist 1979-1983. Atikokan Geological Survey. Open File Report.

Ontario's Mineral Wealth, pamphlet in full color available from the Ontario Ministry of Northern Development and Mines. Good introduction to mining in Ontario. There are excellent pictures and a map.

Prehistoric Copper Mining in the Lake Superior Region, edited and published privately by Professor Roy Ward Drier and Octave Joseph Du Temple, Hinsdale, Illinois. 1961. A series of articles about the ancient peoples who mined copper on Lake Superior.

Open File Report 85-5, Some Questions and Answers on Gold in Wisconsin with a listing of Known Gold Occurrences. W.P. Scott 1985. University of Wisconsin-Extension. Geological and Natural History Survey, 3817 Mineral Point Road, Madison, WI 53705. A short summary of gold occurrences in Wisconsin along with tips and cautions.

Lake Superior Travel Guide Publisher: James Marshall, Lake Superior Magazine, P.O. Box 16417, Duluth, MN 55816-0417. An annual updated booklet of restaurants, resorts, accommodations, attractions, etc. Newest editions can be purchased in many stores around the Lake.

Exploring St. Louis County's Historical Sites. Charles E. Aguar, St. Louis County Historical Society, 1971. Duluth, Minnesota. Very detailed compilation of facts, notes and rumors about almost every township in St. Louis County, Minnesota.

Ghost Mines of the Ely Area, (1882-1925) Milt Stenlund, Ely-Winton Historical Society, 1900 East Camp Street, Ely, MN 55731. Background on the iron and gold mining work near Ely.

A Trace of Gold—The Early Years of Prospecting, by Clem Downey published by Audrey Ferguson, 1985. A book of remembering the early years of gold prospecting in Ontario.

The McKellar Story, McKellar Pioneers in Lake Superior's Mineral Country 1839-1929. E. Marion Henderson published by The McKellar Story Publication Committee, Thunder Bay. Historical book of early pioneers in the Thunder Bay area.

Summer Gold, A Camper's Guide to Amateur Prospecting. John Dwyer, 1971. St. Cloud. A general guide to prospecting around the United States.

Geology and Scenery, Rainy Lake and East to Lake Superior by Dr. E.G. Pye, 1968. Ontario Ministry of Northern Development and Mines (Originally Ontario Department of Mines). A very good guide book to the western part of the region. Many pictures, maps and details. Its companion book, "Geology and Scenery, Pigeon River to Sault Ste. Marie" is out of print. It can be read at the Resident Geologist Office in Sault Ste. Marie or Thunder Bay.

The Mineralogy of Michigan, E. Wm. Heinrich. Published by the State of Michigan CL ' 48 s.321.6. 1976. Available from Publications Room, Geological Survey Division, Department of Natural Resources, P.O. Box 30028, Lansing, Michigan 48909. A nice run through of the geology of Michigan including special features such as clays and gems. Of great help is the alphabetized listing of minerals and gems and known locations where they have been found. A very good book to have for gold or general rock hounding in Michigan.

Gold in Michigan — Open File Report MGSD OFR GOLD 80-1. References and Photocopies of Selected Out-of-print Publications of the Michigan DNR Geological Survey Division, 1980. This is a fascinating compilation of various private reports, notes and newspaper articles about gold finds in Michigan. Many specific locations in the field notes. Available from the Publications Room, Geological Survey Division, Department of Natural Resources, P.O. Box 30028, Lansing, Michigan 48909.

Maps

Public Information Centre, Ministry of Northern Development and Mines, 99 Wellesley St. W., Room 1640, Toronto, Ontario, M7A 1W3. Catalog has many maps (Geological, topographical, others, air photos.) The Centre has access to many additional maps, services and information.

Minnesota Department of Natural Resources, Minerals Division, 1525 East Third Street, Hibbing, Minnesota 55746. (218) 262-6767. This division has maps, location information and core samples available through their Public Land Survey System. This system is a computer data base which can produce all recorded information about any township in Minnesota. In some instances it can produce details for smaller areas down to forty acre parcels. You need to know the township coordinates. You are welcome to stop in at the Hibbing address or call for information.

Ministry of Northern Development and Mines, Mines and Minerals Division, 435 South James Street, Thunder Bay, Ontario P7C 5G6 (807) 475-1331. They have access to all the printed reports and maps available through the government printing office. The Ministry also has an excellent program called the Resident Geologist Program. Trained geologists are available in different parts of the region. They offer technical assistance to prospectors and prospecting classes to the interested public. Call or write the above address for additional information. They also publish "Call on Us!" a short directory of names, addresses and telephone numbers of ministry geologists.

List of Publications in Print, Minnesota Geological Survey, 2642 University Avenue, St. Paul, Minnesota 55114-1057. Ask for current list. Many maps and reports listed with price (most very reasonable).

Canada Map Company, 211 Yonge Street, Toronto, Ont. M5B 1M4, telephone (416) 362-9297 - Fax (416) 362-9381. A licensed distributor for government reports and maps.

Trygg Land Office, P.O. Box 628, Ely, MN 55731, (218) 365-5177. The Trygg maps are compilations of old surveyor maps of Minnesota, Wisconsin, and Michigan including information from the middle to late 1800s. The maps show Indian camps, forest trails, mining locations, and other fascinating details.

Assaying Firms

The following firms are only some of those available to prospectors. This list does not constitute an endorsement of any company or a slighting of any company. It is provided to give you a starting place. Call or write before sending any material. Confirm the services you will need and their cost.

XRAL X-Ray Assay Laboratories, 1885 Leslie Street, Don Mills (Toronto) Ontario M3B 3J4, (416) 445-5755. From U.S. 800-387-0255. Minimum charge $40.00(C). They also have a U.S. subsidiary in Michigan.

Paul's Custom Fire Assaying Ltd., Box 253, Cochenour, Ontario, P0V 1L0 (807) 662-8171.

ACME Analytical Laboratories, Ltd., 852 East Hastings Street, Vancouver, B.C. V6A 1R6, (604) 253-3158, FAX (604) 253-1716.

Skyline Labs, Inc., Denver, CO (303) 424-7718.

Activation Laboratories, Ltd. (ACTLABS), 1336 Sandhill Drive, Ancaster, Ontario L9G 4V5, (416) 648-9611, FAX (416) 648-9613.

International Plasma Laboratory, Ltd. (IPL), 2036 Columbia Street, Vancouver, BC V5Y 3E1, (604) 879-7878, FAX (604) 879-7898. They also have a sample preparation lab in Blaine, Washington.

The Resident Geologist Program

Ontario Ministry of Northern Development and Mines in Thunder Bay District, 435 James St. S., Thunder Bay, Ontario P7E 6E3, (807) 475-1311.

In Kenora District, Box 5200, 808 Robertson St., Kenora, Ontario P9N 3X9, (807) 468-8492.

In Sault Ste. Marie District, 875 Queen St. E., Sault Ste. Marie, Ontario P6A 2B3, (705) 949-1231.

Natural Resources and Research Institute, Center for Applied Research and Technology Development, 5013 Miller Trunk Highway, Duluth, MN 55811-1442, (218) 720-4294.

Prospecting Associations and Newsletters

These are associations of prospectors who exchange information and work for more favorable regulations and laws concerning mineral work.

Prospectors and Developers Association of Canada, 74 Victoria Street, Suite

1002, Toronto, Ontario M5C 2A5. ("PDAC represents the concerns of the Canadian exploration and mining industry to all levels of government." Annual Review 1988-1989).

Canadian Mining Life & Exploration News, Directories North Publications, 35 Birch Street N., Timmons, Ontario P4N 6C8. "Published in the heart of Canada's mining country." About $25.00 for 12 issues.

The Minnesota Prospector, Publication of the Minnesota Exploration Association, 740 East Superior Street, Duluth, MN 55802. The purpose of this group is "to affect public policy regarding non-ferrous exploration and mining in Minnesota in a manner beneficial to its members, and to act as an information clearinghouse for its members." Fee for joining this group is $75.00 (US).

Chambers of Commerce around Lake Superior

You may want to write to one where you will be prospecting for accommodation/camping/equipment information.

Ashland Area Chamber of Commerce
320 Fourth Avenue West
P.O. Box 746
Ashland, WI 54880

Wawa and District Chamber of Commerce
P.O. Box 858
Wawa, Ontario Canada P0S 1K0

Thunder Bay Chamber of Commerce
857 North May Street
P.O. Box 2000
Thunder Bay, Ontario
Canada P7C 4Y4

Sault Ste. Marie Chamber of Commerce
360 Great Northern Road
Sault Ste. Marie, Ontario
Canada P6B 4Z7

Keweenaw Peninsula Chamber of Commerce
326 Shelden Avenue
P.O. Box 336
Houghton, MI 49931

Ontonagon County Chamber of Commerce
P.O. Box 266
Ontonagon, MI 49953

Petoskey Region Chamber of Commerce
401 East Mitchell Drive
Petoskey, MI 49770

Alger Chamber of Commerce
P.O. Box 405
Munising, MI 49862

Sault Area Chamber of Commerce
2581 I-75 Business Spur
Sault Ste. Marie, MI 49783

Marquette Area Chamber of Commerce
501 South Front Street
Marquette, MI 49855

Ironwood Area Chamber of Commerce
100 East Aurora Street
Ironwood, MI 49938

Washburn Area Chamber of Commerce
P.O. Box 638
Washburn, WI 54891

Bayfield Chamber of Commerce
P.O. Box 138
42 South Broad Street
Bayfield, WI 54814

Superior-Douglas County Chamber
　　of Commerce
305 East Second Street
Superior, WI 54880

Two Harbors Area Chamber of
　　Commerce
P.O. Box 39
Two Harbors, MN 55616

Silver Bay Area Chamber of
　　Commerce
P.O. Box 26
Silver Bay, MN 55614

Duluth Area Chamber of Commerce
118 East Superior Street
Duluth, MN 55802

Glossary

Adit. The opening or mouth of a mine.

Amethyst. A variation of quartz in various shades of purple. Amethyst is found in the Thunder Bay area.

Argentite. A common mineral of silver ore.

Basalt. A dark-colored lava with a low silica content.

Claim. A legal marking allowing one person or group the rights to mineral exploration and denying entrance to others.

Concretions. Round balls of sand or clay formed by the rolling action of waves on fine material. Concretions can be found near Ironwood, Michigan, and near Thunder Bay, Ontario.

Crystal. The shape or edges of minerals. Factors such as the crystal chemistry or the rate and spread of crystalization may influence how a crystal is formed. These may make crystals of the same mineral longer, stubbier or banded. Crystals may grow as a pure mineral or as an aggregate.

Crystal faces. The edges or bonding surfaces of a mineral. These form from a solution, melt, or vapor under just the right conditions of temperature and space. Crystal faces result from and conform to the three-dimensional arrangements of atoms. The angles of the faces of any specific mineral sample will be the same when compared to another sample of the same mineral.

Erratic. A piece of rock moved by glaciers and different from the bedrock of an area.

Fault. A fracture where plates of rock have moved against each other.

Greenstone. An altered basic igneous rock, usually basalt, that has chlorite or other minerals that make it green.

Habit. Minerals and mineral aggregates display certain characteristics referred to as their "habits," their general appearances. One important habit is the crystal face. A single crystal will always maintain constant angles for specific forms of minerals. For example, quartz crystals will not look the same as diamonds.

Igneous rocks. Those rocks formed by the crystallization of magma under intense heat, hence the melting or igniting of the rock. Igneous, metamorphic and sedimentary are three major classes of rock found throughout the region.

Intrusion. A body of molten rock that has entered existing rock.

Magma. A body of molten rock formed within the Earth commonly reaching the surface as lava in volcanos.

Metamorphic. Rock that has undergone change (a metamorphosis) due to compression or heat or both.

Occurrence. A location where a mineral has been found in trace amounts.

Prospect. A location where a mineral occurrence has definitely been established and where additional work is warranted to determine the quantity of the mineral.

Quartz. A silica mineral often found in veins and along faults.

Sedimentary. Rock created by the compression of eroded particles, like sandstone.

Serpentine. A greenish silica mineral at times found with gold intrusions.

Striae. Tiny, parallel scratches or grooves found on exposed bedrock and caused by the movement of glaciers.

Vesicle/Vug. Small cavities in lava formed by gas bubbles in cooling molten magma. Agates, amethyst, and other crystals form in vesicles and vugs.